Key Concepts in Computer Science

Key Concepts in Computer Science

Edited by:

Adele Kuzmiakova

ARCLER
PRESS

www.arclerpress.com

Key Concepts in Computer Science

Adele Kuzmiakova

Arcler Press
224 Shoreacres Road
Burlington, ON L7L 2H2
Canada
www.arclerpress.com
Email: orders@arclereducation.com

ISBN: 978-1-77469-152-6 (Hardcover)

ABOUT THE EDITOR

Adele Kuzmiakova is a computational engineer focusing on solving problems in machine learning, deep learning, and computer vision. Adele attended Cornell University in New York, United States for her undergraduate studies. She studied engineering with a focus on applied math. While at Cornell, she developed close relationships with professors, which enabled her to get involved in academic research to get hands-on experience with solving computational problems. She was also selected to be Accel Roundtable on Entrepreneurship Education (REE) Fellow at Stanford University and spent 3 months working on entrepreneurship projects to get a taste of entrepreneurship and high-growth ventures in engineering and life sciences. The program culminated in giving a presentation on the startup technology and was judged by Stanford faculty and entrepreneurship experts in Silicon Valley. After graduating from Cornell, Adele worked as a data scientist at Swiss Federal Institute of Technology in Lausanne, Switzerland where she focused on developing algorithms and graphical models to analyze chemical pathways in the atmosphere. Adele also pursued graduate studies at Stanford University in the United States where she entered as a recipient of American Association of University Women International Fellowship. The Fellowship enabled her to focus on tackling important research problems in machine learning and computer vision. Some research problems she worked on at Stanford include detecting air pollution from outdoor public webcam images. Specifically, she modified and set up a variety of pre-trained architectures, such as DehazeNet, VGG, and ResNet, on public webcam images to evaluate their ability to predict air quality based on the degree of haze on pictures. Other deep learning problems Adele worked on include investigating the promise of second-order optimizers in deep learning and using neural networks to predict sequences of data in energy consumption. Adele also places an emphasis on continual education and served as a Student Leader in PyTorch scholarship challenge organized by Udacity. Her roles as the Student Leader were helping students debug their code to train neural networks with PyTorch and providing mentorship on technical and career aspects. Her hobbies include skiing, playing tennis, cooking, and meeting new people.

TABLE OF CONTENTS

LIST OF FIGURES

LIST OF ABBREVIATIONS

AACs	Advanced Audio Coding
Abend	Abnormal End
ACL	Access Control List
ACPA	Anti-Cybersquatting Consumer Protection Act
ADC	Analog to Digital Converter
ADF	Automatic Document Feeder
ADSL	Asymmetric Digital Subscriber Line
AGP	Accelerated Graphics Port
AIFF	Audio Interchange File Format
AIX	Advance Interactive Executive
ANSI	American National Standards Institute
API	Application Program Interface
ARP	Address Resolution Protocol
ASCII	American Standard Code for Information Interchange
ASP	Active Server Pages
ATA	Advanced Technology Attachment
ATM	Asynchronous Transfer Mode
AV	Anti-virus
BASIC	Beginner's All-purpose Symbolic Instruction Code
BCC	Blind carbon copy
BIOS	Basic Input/Output System
Blob	Binary Large Object
BMP	Bitmap
BPS	Bits Per Second
CCD	Charge-Coupled Device
CLOB	Character Large Object
CMD	Command

CMYK	Cyan Magenta Yellow Black
COM	Component Object Model
CRT	Cathode Ray Tubes
CSS File	Cascading Style Sheets File
Ctrl	Control
DAC	Digital to Analog Converter
DBMS	Database Management Systems
DIMM	Double In-Line Memory Module
DMG	Data Management Group
DNS	Domain Name System
DRAM	Dynamic RAM
DTD	Document Type Definition
DVI	Digital Visual Interface
EEPROM	electronically erasable programmable read-only memory
EPP	Enhanced Parallel Port
FTP	File Transfer Protocol
GHz	Gigahertz
GIF	Graphics Interchange Format
GPT	GUID Partition Table
GUI	Graphic User Interface
HTML	Hypertext Markup Language
IDE	Integrated Drive Electronics
IP	Internet Protocol
ISP	Internet Service Providers
JPEG	Joint Photographic Experts Group
Kbps	Kilobytes per-second
LAN	Local Area Network
MANET	Mobile Ad Hoc Network
Mbps	Megabytes Per-Second
MBR	Master Boot Record
MCA	Micro Channel Architecture
MIDI	Musical Instrument Digital Interface
MOODLE	Modular Object-Oriented Dynamic Learning Environment
MPEG	Moving Picture Experts Group

MySQL	My Sequel
NTFS	New Technology File System
ODBC	Open Database Connectivity
OLAP	Online Analytical Processing
OLE	Object Linking and Embedding
OMG	Object Management Group
OOP	Object-Oriented Programming
OpenGL	Open Graphics Library
OS	Operating System
OSD	On Screen Display
OSPF	Open Shortest Path First
P2P	Peer-to-Peer
PATA	Parallel Advanced Technology Attachment
PCI	Peripheral Component Interconnect
PDF	Portable Document Format
Perl	Practical Extraction and Report Language
PHP	Hypertext Pre-processor
PNG	Portable Network Graphics
PSD	Photoshop Document
QBE	Query by Example
RAM	Random Access Memory
RFID	Radio-Frequency Identification
RGB	Red Green Blue
ROM	Read-Only Memory
SATA	Serial Advanced Technology Attachment
SGML	Standard Generalized Markup Language
SIMM	Single In-Line Memory Module
SLA	Software License Agreement
SMTP	Simple Mail Transfer Protocol
SQL	Structured Query Language
SRAM	Static Random-Access Memory
TCP/IP	Transmission Control Protocol/Internet Protocol
TDMA	Time-division multiple access
TIFF	Tagged Image File Format

UDP	User Datagram Protocol
UML	Unified Modeling Language
UPS	Uninterruptible Power Supply
URL	Uniform Resource Locator
USB	Universal Serial Bus
VANET	Vehicular Ad Hoc Network
VCI	Virtual Channel Identifier
VGA	Video Graphics Array
VHS Tape	Video Home System Tape
VoIP	Voice Over Internet Protocol
VPI	Virtual Path Indicator
VRAM	Video Random Access Memory
WAN	Wide Area Network
WMV	Windows Media Video
XHTML	Extensible Hypertext Markup Language
XML	Extensible Markup Language

PREFACE

Computer science deals with the study of algorithmic processes, computers, and computational systems. As a discipline, computer science spans a wide range of concepts from theoretical studies of algorithms, computation, and information to the practical issues of executing computational systems inside the software and hardware.

Currently, computer science is one of the fastest growing and highest paying career paths in the world. It is considered as a key part of a family of five interrelated disciplines: computer engineering, computer science, information technology, information systems, and software engineering.

Theoretical and practical disciplines are the two major fields of computer science. For instance, the theory of computation deals with abstract models of computation along with general classes of problems that can be resolved using them, while, on the other hand computer graphics or computational geometry highlight more specific applications. Data structures and algorithms have been called the heart of computer science.

Computer science developed as an independent discipline during the early 1960s. However, the electronic digital computer, which is the object of its study, was developed 20 years prior to that. The roots of computer science lie mainly in the related fields of electrical engineering, physics, mathematics, and management information systems.

Programming language theory considers different kinds of approaches to computational processes. Computer programming, on the other hand, involves applying these processes to develop complex systems. Computer architecture defines construction of computer components as well as computer-operated equipment. Artificial intelligence aims to synthesize goal-orientated processes, such as problem-solving, environmental adaptation, decision-making, planning, and learning to predict new data or trends.

The ultimate concern of computer science is determining what can and what cannot be automated. Computer scientists generally emphasize academic research. The key aspect of computer science is problem solving, which is an essential skill for life. People study design, development, and analysis of software as well as hardware that further used to solve problems in a variety of business, social and scientific contexts. As computers solve problems to serve people, there is a substantial human side to computer science as well.

Computer science deals with the process of solving problems. It is also regarded as a field that offers numerous career opportunities. Computer science helps the society by improving healthcare, advancing education sector, or building digital skills.

In the modern world, computer science can predict and avoid disasters. Every tool used to monitor weather through satellites uses radar, which is built using technology. Using those tools, computer science allows to forecast hurricanes, tornadoes, or even the outbreak of a fatal disease Computer science solves problems that can help people everywhere in the world. According to a U.S. News & World Report, earning a graduate degree in computer science can eventually lead to positions in some of the top companies in the world.

This book will introduce the readers to the field of computer science and the fundamentals of computer programming. It is precisely designed for readers or students with no prior programming experience, and touches upon a diversity of fundamental topics. By the end of the book, readers will understand the basics of computer science and the key concepts involved.

A

Abend

The term abend is abbreviated as "abnormal end." An abend occurs when a process comes to an unexpected or abnormal end. When it comes to computer software, the term abend usually refers to a software crash, which occurs when a program unexpectedly shuts down. A bug in a program's code could possibly cause it to freeze or crash when running a command. As a result, the program comes to an abrupt (and frequently inconvenient) finish. Initially the term "ABEND" was used as an error message by IBM OS/360 systems. In the modern world, it is commonly used by Novell Netware systems and is also used as a general programming term.

Access

Microsoft Access is a popular database application for Windows that is sometimes shortened as "MS Access." Users can use Access to create databases that store data in a logical order. A visual interface is also provided for designing custom forms, tables, and SQL queries. Visual forms or a rudimentary spreadsheet interface can be used to enter data into an Access database. Other programs, including Web services, can browse, search, and access the information stored in an Access database.

While Access is a proprietary database management system (DBMS), it supports Open Database Connectivity, making it interoperable with other database systems (ODBC). Data can be transported to and from other database programs, such as MS SQL, FoxPro, Filemaker Pro, and Oracle databases, using this method. Because of this interoperability, Access can also be used as the back end for a database-driven website. In fact, Access databases are supported by Microsoft FrontPage and Expression Web, as well as ASP.NET. As a result, access databases are frequently used to generate dynamic content on websites based on Microsoft Windows servers.

Access Point

An access point is known as a device that allows users to connect to a network wirelessly. Access point-connected devices can communicate with other network devices. If the access point is connected to an Internet connection, as is usually the case, they can connect to the Internet as well.

Wi-Fi access points are also known as base stations. For example: "The coffee shop provides an access point for customers with Wi-Fi devices."

ACL

ACL is the abbreviation for "Access Control List." For a file, folder, or other object, an ACL is a list of user permissions. It specifies which users and groups are permitted to access the item and what operations they are permitted to do. Read, write, and execute are common examples of these operations. For instance, if an ACL grants read-only access to a file to a certain user, that user will be able to open the file but not write to it or run it.

The use of access control lists to manage file and folder permissions is simple. Most operating systems, including Windows, Mac, and Unix, make use of them. ACLs are usually hidden from the user, although they can easily be changed via a graphical interface. In Mac OS X 10.5, for example, ACL settings can be changed by clicking a file or folder and selecting "Get Info." The access control settings can be changed under the window's "Sharing and Permissions" section. The chmod() command on Unix systems can be used to change ACLs.

Active-Matrix

High-quality flat-panel displays, such as laptop screens and thin computer monitors, use active-matrix technology. Diodes, or miniature semiconductors, are laid over a grid of ultra-fine wires to form the images on active matrix panels. When current runs through the diodes, it causes them to light up in various colors, depending on the current's strength. The image on the screen is formed by thousands of these diodes placed next to each other.

Active-matrix displays employ transistors to keep the diodes on or off, which aren't found in lower-quality passive-matrix displays. Active-matrix displays are brighter and have higher contrast than passive-matrix displays thanks to the transistors.

ActiveX

This Microsoft-developed technology was created to connect desktop applications to the Internet. Software developers can use ActiveX development tools to build interactive Web content for their apps. Word and Excel documents, for example, can be viewed natively in ActiveX-enabled Web browsers. While ActiveX is a helpful technology, using ActiveX-enabled material requires that you have an up-to-date version of ActiveX installed on your PC.

Ad Hoc Network

The term "ad hoc" is derived from the Latin phrase "for this purpose." It is a term that is frequently used to describe solutions that are created on the spot for a specific goal. An ad hoc network is a network connection that is established for a single session and does not require the use of a router or a wireless base station in computer networking.

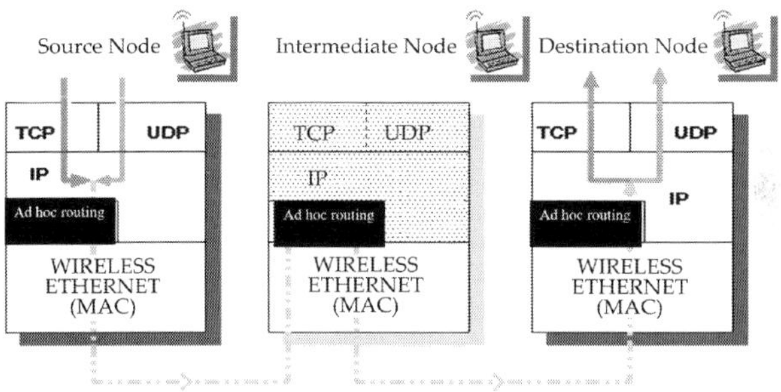

Figure 1. Ad Hoc network.

Source: Image by Wikimedia.

If you need to transmit a file to your friend's laptop, for example, you may set up an ad hoc network between your computer and his laptop. This can be accomplished by using an Ethernet crossover connection or the wireless cards on the PCs to communicate with one another. Set up a multi-hop ad hoc network if you need to share files with more than one computer.

An ad hoc network is essentially a transient network connection made for a specific purpose (such as transferring data from one computer to another). It is a regular local area network (LAN) if the network is set up for a longer period of time.

ADC

ADC is the abbreviation for "Analog-to-Digital Converter." Computers can only process digital data; hence they need digital input. As a result, an analog-to-digital converter (ADC) is required when an analog input is transferred to a computer. This device can digitize an analog signal, such as an electrical current, into a binary representation that can be understood by a computer.

The conversion of analog video to digital format is a popular use for an ADC. Video captured on 8 mm film or a VHS cassette, for example, is saved in an analog format. The video must be converted to a digital format before it can be transferred to a computer. An ADC video converter box, which normally has composite video inputs and a Firewire output, can be used to accomplish this. Some digital camcorders have analog inputs that can be used to convert video from analog to digital.

Analog audio streams can also be converted using ADCs. If you wish to record sounds from a microphone, for example, you will need to convert the audio from the microphone's analog signal to a digital signal that the computer can comprehend. This is why all sound cards with an analog audio input also need an ADC to transform the signal to a digital representation. The sample rate employed in the conversion process determines the accuracy of the audio conversion. Enhanced sampling rates allow for a more accurate assessment of the analog signal, resulting in higher sound quality.

While ADCs transform analog inputs into a digital format that computers understand, a computer may occasionally need to produce an analog signal. A digital-to-analog converter (DAC) is used for this type of conversion. Important: ADC can alternatively stand for "Apple Display Connector," an Apple-developed proprietary video connector. It was a single cable that combined DVI, USB, and AC power. In 2004, Apple discontinued making computers with ADC connections in favor of the more common DVI connection.

Add-on

An add-on is a software expansion that enhances a program's functionality. It may add new objects to the program's interface, extend particular functions within the program, or offer the software new capabilities. Add-ons such as the Google toolbar, ad blockers, and Web developer tools, for example, are supported by Mozilla Firefox, a popular Web browser. Some computer games allow players to purchase add-ons that include extra maps, new characters, or the ability to customize the game.

The majority of add-ons come in self-installing packages. This implies that the user may just double-click the add-on package to install the program's files. Other add-ons may need the user to manually move files into specific directories. While not all applications enable add-ons, many programs are now designed with add-on support since it allows other developers to easily extend the program's functionality.

These supplementary functions are not always referred to as "add-ons" in all software applications. Dreamweaver, for example, provides "extensions," which add new Web development functions, whereas Excel can import "Add-Ins," which add extra spreadsheet tools to the user. Plug-ins, which are a form of add-on, are also supported by many programs.

ADF

ADF is the abbreviation for "Automatic Document Feeder." In copy machines and scanners, an ADF is used to feed papers into the machine. It lets you copy or scan numerous pages at once without having to set each one in the copier or scanner separately.

A flatbed scanner, on the other hand, necessitates the placement of each sheet on the scanning surface individually. While this is OK for single pages, copying 80 would necessitate a lot of manual page loading! Some copy machines come equipped with both an automatic document feeder and a flatbed scanning surface, allowing you to use either approach for copying documents.

ADSL

ADSL is the abbreviation for "Asymmetric Digital Subscriber Line." ADSL is a sort of digital subscriber line (DSL), which is a technique of sending data over copper phone lines. While symmetrical DSL (SDSL) uploads and downloads data at the same speed, ADSL uploads and downloads data at varying maximum data transfer rates.

An ADSL connection, for example, may support download speeds of 1.5Mbps while upload speeds are limited to 256Kbps. Because most users download far more data than they upload, the difference between the two rarely affects Internet connection speeds. However, ADSL would be inefficient for Web servers or other systems that transmit a lot of data upstream.

Adware

Adware is a type of free software that is monetized through adverts. Toolbars that sit on your desktop or work in combination with your Web browser are examples of common adware programs. They offer capabilities such as improved Web or hard drive searching, as well as superior bookmark and shortcut organization. Adware can sometimes take the form of more complex programs like games or utilities. They are free to use, but you must view adverts for the duration of the program. Adware often requires an active Internet connection to execute because the adverts often allow you to click to a Web site.

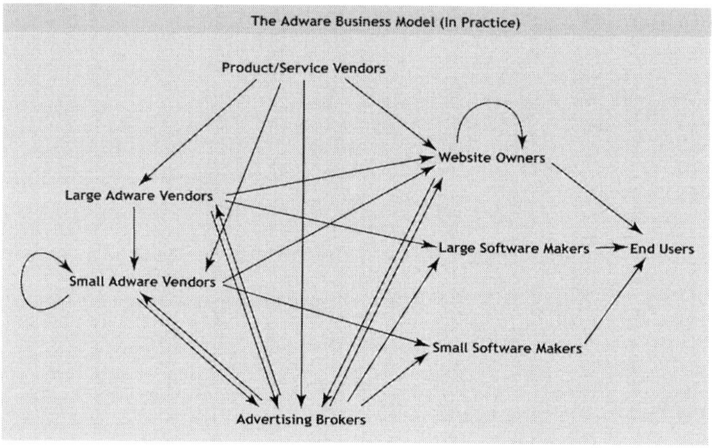

Figure 2. Adware.

Source: Image by Flickr.

Although most adware is harmless, some can act as spyware, collecting information about you from your hard drive, Websites you visit, or keystrokes. Spyware applications can then communicate the information to another machine through the Internet. As a result, one must be cautious about what adware they install on their computer systems.

Affiliate

Internet marketing is propelled by website affiliates. Affiliate programs are used by businesses to earn leads and sales from other websites. They pay a commission to the websites that host their adverts for things sold through the links on their pages. If a site owner joins Amazon's affiliate program, for example, he will receive ad banners or links from Amazon that he can display anyplace on his website. If a visitor purchases something after clicking on the Amazon.com banner or a link on his site, he will receive a commission. Unfortunately for website owners, affiliate commissions rarely exceed 5% because most online sales have low profit margins.

In the late 1990s, the development of Website affiliate programs catapulted Internet marketing to new heights. The first companies to provide these programs witnessed a significant rise in revenue. After all, these businesses essentially receive free promotion and pay their affiliates a percentage of the sales they create. For the affiliate program's operator, it's a win-win situation. The only disadvantage is that setting up and maintaining the programs takes a lot of time (and often a lot of money). As a result, smaller businesses may not see the value in having an affiliate program.

AGP

AGP is the abbreviation for "Accelerated Graphics Port." This is a graphics card expansion port on a computer's motherboard that was designed by Intel. PCI graphics connections have a maximum transfer rate of 132 MB/sec and normally run at 33 MHz AGP ports, on the other hand, operate at 66 MHz and can transport data at speeds of up to 528 megabits per second. This allows games and apps to store and retrieve larger, more realistic 3D shapes and textures without causing the animation on the screen to slow down. AGP cards can also store graphics in system memory rather than visual memory, which in turn helps with performance. AGP graphics cards will often have better performance per MB of VRAM than PCI graphics cards due to these advantages.

AIFF

AIFF is the abbreviation for "Audio Interchange File Format." AIFF is an audio format for storing high-quality sampled audio data that was created by Apple Computer.

Figure 3. Audio interchange file format.

Source: Image by Thenounproject.com.

In terms of size and quality, AIFF files are equivalent to Windows WAVE files. AIFF and WAVE files can both store CD-quality audio and can be burned on an audio CD. Even though Apple designed the AIFF format, audio apps on both the Mac and PC can usually read the files.

AIX

Although some Linux enthusiasts have referred to it as "Ain't UNIX," it stands for "Advanced Interactive Executive." IBM's AIX operating system is based on the Unix operating system. It is designed for enterprise servers and has a comprehensive set of security features like Kerberos V5 network authentication and dynamic secure tunnel authentication. The system administrator can divide memory, CPU, and disc access amongst different jobs in AIX. The system is backwards compatible with 32-bit software and supports IBM's 64-bit POWER processor. It also supports Java 2 and can run most Linux applications (after recompiling them). If you are confused, don't worry: AIX isn't your ordinary consumer operating system. It is primarily used in huge corporations for servers, where IT nerds get to work with it.

Alert Box

An alert box, also known as a message box, is a little window that appears on your screen to notify you when your computer is about to conduct a possibly harmful operation. When you empty your computer's trash or recycle bin, for example, an alert box may appear that asks, "Are you sure you want to permanently erase these items?" After that, you have the option of selecting "OK" to delete the items or "Cancel" to prevent them from being deleted.

Users are protected by alert boxes, which prevent us from doing something people regret. The notice box that appears when you try to close a document without saving it is probably the most common. "Save changes to this document before closing?" will most likely appear in an alert box. You can choose between "Don't Save," which will reject the changes, "Save," which will save the changes before closing the document, and "Cancel," which will prevent the document from closing and allow you to continue working on it.

To catch your attention, most alert boxes use the standard alert icon, which is a triangle with an exclamation point in the middle. However, not all alert boxes, such as "Cancel" and "OK," contain several possibilities. An alert box may appear on your screen, for example, stating that a software has unexpectedly quit after performing an unauthorized operation. When this happens, your only alternative is to hit "OK" and then smack your computer in the face for leaving the software before you could save your work.

Algorithm

An algorithm is a set of instructions, sometimes known as a method or a function, that are used to complete a certain task. This might be as basic as putting two integers together or as sophisticated as applying effects to an image. To sharpen a digital photo, for example, the algorithm would have to process each pixel in the image and determine which ones to change and by how much such change is to be affected to make the image look sharper.

The majority of computer programmers devote a significant portion of their work to developing algorithms. (They spend the rest of their time debugging algorithms that aren't working properly.) The goal is to develop efficient algorithms that do not waste unnecessary computer resources (such as RAM and CPU time). This can be challenging since a method that works well on one piece of data may not work as well on another.

Poorly constructed algorithms, as you might expect, can cause applications to operate slowly or even crash. As a result, software upgrades are frequently released with claims of "enhanced stability and performance." This is impressive, but it also suggests that prior versions of the software's algorithms were not written as properly as they could have been.

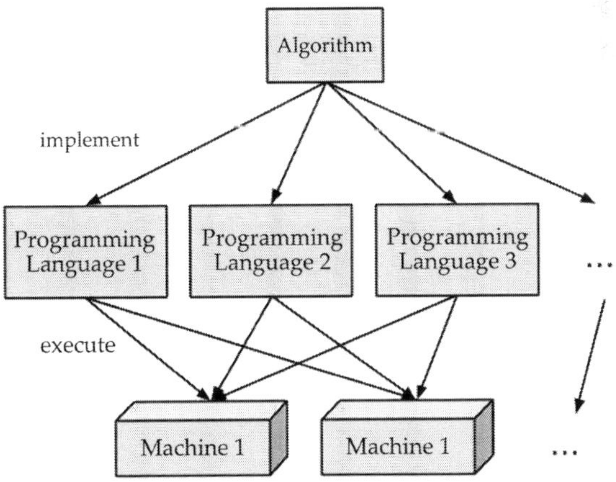

Figure 4. Algorithm.

Source: Image by Wikimedia.

Analog

As humans, they have an analog perception of the world. Everything they see and hear is a constant stream of data reaching our senses. Analog data is defined by this continuous stream. Digital information, on the other hand, uses only ones and zeros to approximate analog data.

A turntable (also known as a record player) is an analog device, whereas a CD player is digital. This is because a turntable reads a continuous signal from a record, whereas a CD player merely reads a series of ones and zeros. A VCR, on the other hand, is an analog device, whereas a DVD player is digital. A VCR reads a tape as a continuous stream of information, but a DVD player only reads ones and zeros from a disc.

Digital devices can only approximate an audio or video stream because they can only read ones and zeros. As a result, analog data is more precise than digital data. Digital data, on the other hand, is easier to alter and preserve than analog data. Furthermore, computers can only handle digital data, which is why most information is now stored digitally. However, you are not out of luck if you wish to transfer video from old analog video tapes to your computer so you may edit them. A digital to analog converter (DAC) can be used to transform analog data into a digital signal that your computer can understand.

Android

Android is a mobile operating system (OS) developed by Google and used by several smartphones, including the G1, Droid, and Nexus One. Android is open source, which means it isn't tied to any particular phone manufacturer or service provider. The Android operating system also supports third-party applications, which can be downloaded from Google's online software store, Android Market. For example: "The Android operating system can be customized for each mobile device."

Figure 5. Android

Source: Image by Pixabay

ANSI

ANSI is the "American National Standards Institute." By developing and facilitating voluntary consensus standards and conformity assessment systems, ANSI aims to promote the global competitiveness of American enterprises and the quality of life in the United States. Rather than being regulated by the government, "voluntary" standards are determined by market requirements. "Conformity assessment systems" assess a product's or service's adherence to a set of guidelines.

ANSI does not create standards or administer examinations; rather, it helps this work by "accrediting" (that is, recognizing) competent bodies with the necessary expertise. The Institute represents the US at the International Organization for Standardization (ISO) and the International Electrotechnical Commission (IEC). The American National Standards Institute (ANSI) is a private, non-profit membership organization with offices in Washington, DC, and New York City. The ANSI Web site is the best place to go if you want to learn more about the organization. More information on ANSI and what they do can be found there, as well as the most recent news on national and international standards.

Antivirus

Antivirus (sometimes spelled "anti-virus" or just "AV") is computer software that prevents, detects, and removes infections and harmful programs. Most antivirus programs work in the background, scanning new files for viruses and running regular system checks once they've been installed. You can also scan individual files or folders with an antivirus program.

The software works by comparing files to a virus definitions database, which contains signatures of known infections. If a file is infected with a virus, the antivirus program will usually quarantine it, rendering it unavailable (and therefore removing the threat). It may also indicate that the file should be deleted. For example: "You should run an antivirus program to make sure your computer isn't infected."

Figure 6. Antivirus.

Source: Image by Pxhere.com.

While antivirus software is primarily used to detect viruses, many modern antivirus applications also scan for other types of malware. Trojan horses, worms, spyware, and adware are all examples of this. Many antivirus products are now incorporated as part of "internet security" suites, which contain features like customized firewalls, website blocking, and port monitoring in addition to antivirus protection.

API

Although it is commonly referred to as an "Application Programming Interface," it stands for "Application Program Interface (API)." A programming interface (API) is a set of commands, functions, and protocols that programmers can use to create software for a particular operating system. The API allows programmers to interface with the operating system by calling predefined functions rather than developing them from scratch.

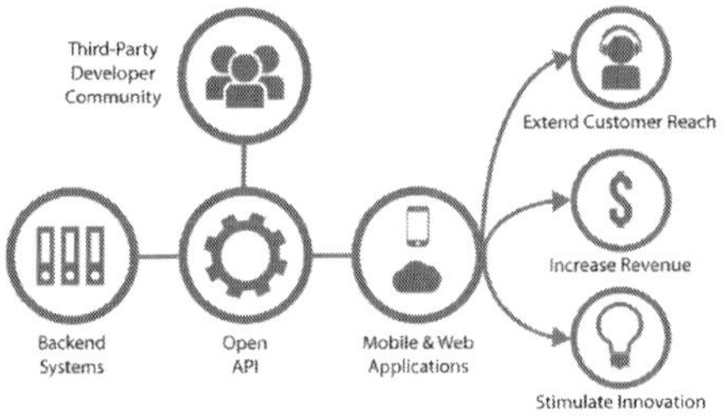

Figure 7. Application Program Interface.

Source: Image by Wikimedia.

For programmers, all computer operating systems, including Windows, Unix, and Mac OS, have an application program interface. Video gaming consoles and other hardware devices that can run software programs employ APIs as well. While the API makes programming easier for programmers, it also benefits the end user by ensuring that all apps that use the same API have a similar user interface.

Apple

Apple, situated in Cupertino, California, is a technological corporation. Steve Jobs, Steve Wozniak, and Ronald Wayne created it in 1976. The iMac, Mac mini, MacBook, MacBook Pro, and Mac Pro are all Macintosh computers made by Apple. Apple also produces several personal electronic gadgets, such as the iPhone, iPad, iPod, and Apple Watch.

Apple is most recognized for its hardware products, but it also creates a wide range of software tools. The macOSOS X operating system, the Safari web browser, and the iWork programs are all examples. Final Cut Pro and Logic Pro are two professional media software developed by Apple.

Apple also provides a number of different subscription platforms, including Tunes Store, App Store, and the upcoming Mac App Store are all examples. You can download media and programs directly to your PC using these services. Apple also has over 300 retail outlets throughout the world that sell Apple products and provide customer service.

Applet

This is a Java application that may be included in a Web page. A Java applet differs from a conventional Java application in that it cannot access system resources on the local computer. The applet cannot access or use system files or serial devices (modems, printers, scanners, and so on). This is for security concerns; no one wants their computer to be wiped clean by a rogue applet on a bad actor. Applets helped to make the Web more dynamic and engaging and gave the Java programming language a boost.

Application

A computer application, often known as an application program, is a piece of software that runs on your computer. Applications include web browsers, email programs, word processors, games, and utilities. Because each software has a unique application for the user, the term "application" is utilized. A word processor, for example, can assist a student in writing a research paper, while a video game can keep the student from getting distracted.

System software, on the other hand, is made up of programs that operate in the background and allow apps to run. Assemblers, compilers, file management tools, and the operating system itself are among these programs. Because the system software is made up of "low-level" programs, applications are said to run on top of it. While system software is loaded along with the operating system, you have the option of selecting which programs to install and run on your computer.

Windows programs are commonly referred to as executable files, whereas Macintosh programs are commonly referred to as apps. This is why Mac programs end in-app and Windows programs end in.EXE. Even though their file extensions are different, Macintosh and Windows programs serve the same purpose and are both classified as apps.

Archie

Archie is a program that searches FTP (File Transfer Protocol) servers for files. Archie isn't used very often anymore because it requires that you to know the exact file name. The majority of file searches are now conducted via the Internet using a Web browser, such as Firefox, Chrome, or Safari.

Archive

An archive is a single file that contains several files and/or folders. Archives can be preserved in a variety of formats, the majority of which use file compression to reduce file size. They are used to assemble a collection of files into a single package. When moving several files to another user or backing up data, this can be useful.

An archive's files must first be extracted using a file extraction program before they can be opened. Before the files can be extracted from a compressed archive, it must first be decompressed. Both of these steps are usually performed automatically by most file decompression systems.

ARP

"Address Resolution Protocol" or ARP is a mechanism that maps an IP address to a computer on a local area network (LAN). The ARP translates the IP address to the MAC address since each machine has a unique physical address called a MAC address. This guarantees that each computer has its own network ID.

When data supplied to a network arrives at the gateway, which acts as the network's entrance point, the Address Resolution Protocol is used. Based on the IP address to which the data is being delivered, the gateway utilizes ARP to locate the computer's MAC address. This information is usually found in a table termed the "ARP cache" by the ARP. If the address is located, the information is passed on to the gateway, which routes the data to the correct computer. If necessary, it can also transform the data to the right network format.

If the address cannot be discovered, the ARP sends out a "request packet" to other machines on the network to determine if the IP address belongs to one of the machines not in the ARP cache. If a genuine system is discovered, the information is conveyed to the gateway, and the ARP cache is updated with the new data. Future requests for that IP address will be substantially faster after you update the ARP cache. While this may appear to be a time-consuming procedure, it normally just takes a fraction of a second to finish. If only finding old receipts when you need them was as simple.

Array

A data structure that contains a group of elements is known as an array. In computer programs, arrays are widely used to organize data so that a related group of numbers can be readily sorted or searched. A search engine, for example, might utilize an array to store Web sites found during a user's search. The software will output one element of the array at a time when showing the results. This can be done for a set number of values or until the entire array of values has been output. While the program might generate a new variable for each result, putting the results in an array is a much more effective method of memory management.

The syntax for storing and displaying the values in an array typically looks something like this:

```
arrayname[0] = "This";
arrayname[1] = "is";
arrayname[2] = "pretty simple.";
print arrayname[0];
print arrayname[1];
print arrayname[2];
```

The above commands would output the array's first three values, or "This is fairly straightforward." The programmer can tell the program to output each value in the array until the last value is reached by using a "while" or "for" loop. As a result, arrays not only assist programmers manage memory more effectively, but they also make the programmer's job easier.

ASCII

Computers utilize ASCII, or American Standard Code for Information Interchange, to represent all upper and lower-case letters, digits, and punctuation. Without ASCII, each type of computer would have its own way of expressing letters and numbers, causing enormous confusion for computer programmers (and allowing them to sleep even less than they already do).

Text can be displayed in the same way on a Dell Dimension in Minneapolis, Minnesota, as it can on an Apple Power Mac in Paris, France, thanks to ASCII. There are 128 ASCII codes, each of which can be represented by a 7-digit binary number (because 2^7 = 128).

ASP

Just to confuse ordinary people like you and me, there are two equally important definitions of ASP. The term can refer to (1) an Active Server Page, or (2) an Application Service Provider.

1. Active Server Page

This is a Web page that contains one or more ASP scripts. When an ASP-based Web page is visited, ASP scripts run like small computer programs. If the URL's suffix is ."asp," you're on an active server page (as opposed to ".html"). Before being sent to a user's browser, ASP pages are processed on a Web server.

ASP pages are commonly used for pages that include dynamic or constantly changing data. An ASP script might obtain a visitor's zip code via a Web form and then customize the content on the resulting page based on that information. ASP scripts are often written in Microsoft's Visual Basic programming language, as ASP technology was created by Microsoft.

2. Application Service Provider

This is a third-party corporation that distributes software-based services from a central site to customers in various locations, sometimes known as an "app-on-tap." Companies can utilize ASPs to outsource services that would otherwise have to be done in-house or on-site. Using an ASP to manage information services is typically a cost-effective option for businesses and organizations. There are five main categories of Application Service Providers:

Local or regional application service provider (ASP) – provides a variety of application services to small enterprises and people in a specific geographic area.

Specialist ASP – develops applications for specific purposes, such as HR or Web services.

- Vertical Market ASP – provides support to a specific industry such as Education.
- Enterprise ASP – provides high-end business with information and services.
- Volume Business ASP – provides high-volume services to small and medium-sized businesses.

ASP.NET

ASP.NET is a set of Web development tools offered by Microsoft. Programs like Visual Studio .NET and Visual Web Developer allow Web developers to create dynamic websites using a visual interface. of course, programmers can write their own code and scripts and incorporate it into ASP.NET websites as well. Though it is often seen as a successor to Microsoft's ASP programming technology, ASP.NET also supports Visual Basic.NET, JScript .NET and open-source languages like Python and Perl.

The ASP.NET framework is based on the.NET framework, which provides software programmers with an application program interface (API). The.NET development tools can be used to create apps for both Windows and the Web. Visual Studio.NET, for example, provides a visual interface for developers to create their apps, making .NET a viable option for developing Web-based applications.

An ASP.NET website must be published to a Web server that supports ASP.NET apps in order to work properly. By far the most popular platform for ASP.NET websites are Microsoft's Internet Information Services (IIS) Web server. While there are various open-source choices for Linux-based computers, these options often fall short of providing complete support for ASP.NET apps.

ATA

"Advanced Technology Attachment" is a disc drive that has the drive controller built inside it. ATA hard drives can be used in computers without the need for a special controller. A separate card (such as a SCSI card for a SCSI hard drive) is not required, but the motherboard must still support an ATA connection. Some different types of ATA standards include ATA-1, ATA-2 (a.k.a. Fast ATA), ATA-3, Ultra ATA (33 MBps maximum transfer rate), ATA/66 (66 MBps), and ATA/100 (100 MBps).

IDE stands for "Integrated Drive Electronics," and it's also how ATA drives are referred to. ATA drives are sometimes labelled as "IDE/ATA" to confuse consumers who are buying hard drives. ATA uses IDE technology technically, but the crucial thing to remember is that they both refer to the same object.

ATM

"Asynchronous Transfer Mode" is the abbreviation for "Asynchronous Transfer Mode." The majority of people are familiar with ATMs as automated teller machines, which are friendly boxes that allow you to take cash from your bank or credit account. ATM, on the other hand, has a different meaning in the computing world. Asynchronous Transfer Mode (ATM) is a networking protocol that sends data in packets or cells.

53-byte cells are used in ATM (5 bytes for the address header and 48 bytes for the data). These incredibly small cells can be processed quickly enough through an ATM switch (not an automated teller machine) to sustain data transfer speeds of over 600 megabits per second. The technology was created to transmit all types of material at fast speeds, from simple graphics to full-motion video.

ATM equipment can transfer massive volumes of data over a single link while ensuring that no single transmission consumes all of the bandwidth since the cells are so small. It also allows Internet Service Providers (ISPs) to assign each client a certain amount of bandwidth. While this may appear to be a disadvantage to the client, it really improves the efficiency of the ISP's Internet connection, resulting in a speedier overall connection for everyone.

Autocomplete

Autocomplete is a text prediction system that works as you type. When you type a query into a search box, for example, autocomplete will show you a list of options that start with the string you typed.

The suggestions supplied by autocomplete are usually based on the search engine's history of user searches. However, information like your location or browser history may be included in the autocomplete algorithm. As a result, the autocomplete suggestions you see may differ from those of others.

Autocomplete serves two main purposes — to give you search suggestions and save you time If you notice a search phrase you wish to use, simply click on it, or use your keyboard's arrows to get to it, then hit Enter.

Autoresponder

A mail server's autoresponder is a program or script that automatically responds to e-mails. An autoresponder can normally be set up by the user through a Web-based interface, even if it is administered by the mail server. For instance, a corporation might set up an autoresponder for its support e-mail address to notify users that their help requests have been received. It is possible that the automated response will say something along those lines: "Thank you for your message; we have received it. After he finishes his dart game in the lobby, one of our techs will try to address your query."

Avatar

An avatar is a digital representation of a person. Players can create bespoke characters, known as avatars, in games like World of Warcraft and The Sims Online, for example. In an online environment, people can control their avatars and communicate with other players. The icon and username that a user picks while registering for a website or online discussion forum can also be referred to as an avatar. The avatar name and image typically show up next to any contributions the user makes on the website. For example: "His World of Warcraft avatar is a Mage with black armor and a silver staff."

B

Backbone

A network backbone transports data to smaller transmission lines, in a similar way that the human backbone transports signals to many smaller neurons throughout the body. The major network cables that connect numerous local area networks (LANs) together are referred to as a local backbone. As a result, a wide area network (WAN) with a backbone connection has been created.

The Internet, as the world's largest wide-area network, relies on a backbone to transport data over long distances. The Internet backbone is made up of multiple ultra-high-bandwidth links that connect a variety of nodes all around the world. Incoming data is routed through these nodes to smaller networks in the area. The fewer "hops" your data must take before reaching the backbone, the faster it will be delivered to its destination. This is why many Web servers and ISPs have direct Internet backbone connections.

Backside Bus

Data is sent to and from a computer's CPU via two types of buses. The front side bus and the backside bus are the two types of buses. Surprisingly, there is no link between them and the backside and frontside airs popular among snowboarders.

The backside bus transports data to and from the computer's secondary cache, while the front side bus transports data between the CPU and memory. The L2 cache, or secondary cache, keeps frequently used functions and data close to the processor. Because it can repeat tasks faster, the computer's CPU can work more efficiently.

The backside bus is used to send information from the L2 cache to the processor. Because this operation must be completed quickly, the rear bus's clock speed cannot be allowed to fall behind. As a result, the backside bus is frequently faster than the processor. On the other hand, the front side bus is usually half the speed of the processor or slower.

Backup

Backup is the most important computer term one should be aware of. A backup is a duplicate of one or more files prepared as a fallback if the original data is lost or becomes unusable. For example, you might keep numerous versions of a research paper as backup files on your hard drive in case you decide to use a previous draught. Better yet, save the backups to a USB flash drive, which will secure the files if the hard drive fails.

Hard drives are designed to last a long time without failing. However, they, like all modern equipment, they are not without flaws. Hard drives rely on moving parts to retrieve data because they are not solid-state devices, which can fail and render your data unrecoverable. Go to your local computer store and have someone show you an open hard drive if you need proof of how delicate hard drives are. When you realize that all your data is saved on such a small, delicate device, you may have a new appreciation for the importance of data backup.

But it's not simply hardware failures to be concerned about. Your files may potentially be harmed by software corruption. Damaged directory structures can result in the disappearance of entire folders. Viruses and other software attacks can corrupt or remove files by accident. Conflicts in program installation might render apps or files unusable. Unfortunately, there are numerous ways for your data to be lost or destroyed.

That is why it is critical to have a backup of your data. Many individuals do not recognize the value of having a backup until it's too late. of course, maintaining a backup becomes even more important when you've lost years of images, school papers, business documents, e-mail archives, music, movies, or any other data that you can't retrieve.

So how do you backup your data? The best way is to use an external storage device, such as an external hard drive, flash memory device, or even another computer. You can also create permanent backups using optical media, such as CD-R and DVD-R discs. Backing up individual folders and files is as easy as copying them from the source media (your computer's hard disk) to the destination (an external hard drive). If you want to back up your entire system or would like to have regular backups automatically performed, you can use backup software that will back up your data for you. Many programs are available for both Mac and Windows that provide automatic backups and system restore capabilities.

Bandwidth

The amount of data you can transport via a network or modem connection is referred to as bandwidth. Bits per second, or "bps," is the most common unit of measurement. Bandwidth can be compared to a highway with cars travelling on it. The roadway represents the data, and the automobiles are the network connection. The more cars that can travel on a roadway at the same time, the broader it is. As a result, more cars will arrive at their destinations faster. In the case of computer data, the more bandwidth available, the more data that can be transported in a given length of time.

Banner Ad

Whether you like it or not, advertising powers most of the Internet. Websites, like television and radio, can provide free content in exchange for advertising revenue. While you may become tired of seeing commercials on the Internet from time to time, most people would agree that seeing a few ads here and there is preferable to paying a monthly fee for each website.

The banner ad is perhaps the most used type of online advertising. It's a long, rectangular graphic that can go nearly anyplace on a Web page. The majority of banner advertisements are 468 pixels wide by 60 pixels tall (468x60). They may feature text, photos, or obnoxious animations that make it difficult to concentrate on the page's content. When a user clicks on a banner ad, he or she is taken to the advertiser's website, regardless of the type of ad.

Base Station

The word "base station" was first used to describe the cell phone towers that can be seen on the side of the road. These stations take care of all cellular calls within their service area, receiving data from one end of the line and transmitting it to the other.

A base station, on the other hand, is the wireless access point for PCs with wireless cards in the computer industry. It is essentially a router that communicates with Wi-Fi-enabled devices. 802.11b and 802.11g are two prevalent Wi-Fi setups. Companies such as Netgear, Linksys, D-Link, Apple Computer, and others produce wireless base stations. Fortunately, all wireless cards can connect with base stations from any manufacturer as long as the hardware is based on the Wi-Fi standard.

Figure 8. Base Station.

Source: Image by Wikimedia.

Baseband

A baseband signal is a signal that has not been modulated or has been demodulated back to its original frequency. To be sent over long distances, most telecommunications protocols require baseband signals to be transformed, or modulated, to a higher frequency. As a result, throughout the transmission process, the original baseband, or lowpass, signals are converted. The signal is demodulated when it reaches its destination, allowing the recipient to receive the original baseband signal. Because it delivers data in baseband, Ethernet is an example of a protocol that does not require signal modulation. For example: "Baseband signals must be modulated to higher frequencies for radio transmission."

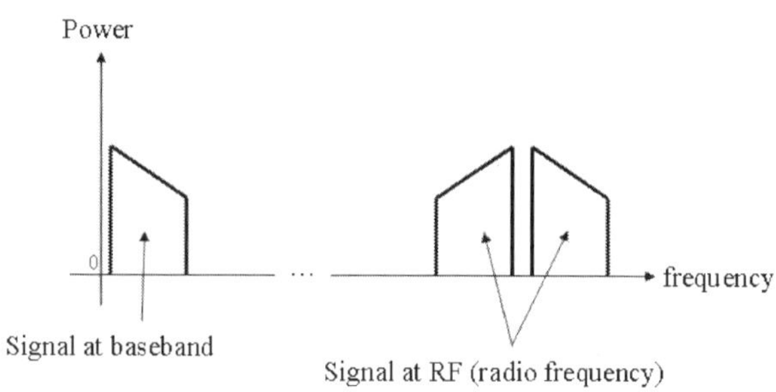

Figure 9. Baseband.

Source: Image by Wikimedia.

BASIC

"Beginner's All-purpose Symbolic Instruction Code" is the acronym for "Beginner's All-purpose Symbolic Instruction Code." BASIC is a computer programming language that was created in the mid-1960s to help students write basic computer programs. Since then, the language has matured into a more sophisticated and powerful programming language that can be used to construct advanced applications for today's computers.

Initially, BASIC used numbers at the beginning of each instruction (or line) to tell the computer what order to process the instructions. Lines would be numbered as 10, 20, 30, etc., which would allow additional instructions to be placed between commands later if needed. "GOTO" statements enabled programs to loop back to earlier instructions during execution. For example, line 230 of a BASIC program may have an "if" clause that tells the computer to jump back to line 50 if a variable is less than 10. This instruction might look something like this:

```
230 IF (N < 10) THEN GOTO 50
```

While loops, which perform a set of instructions as long as a specific condition is true, are used in more contemporary BASIC implementations. For storing variables and other data, newer BASIC programming software offers more data types, such as integers, strings, and arrays. While the early BASIC development environments were solely text-based, today's BASIC programming software allows programmers to create most of their code visually using a graphical user interface. REAL basic and Microsoft Visual Basic are two of the most popular BASIC development programs today.

Batch File

A batch file is a script that has a list of commands in it. These commands are run in a specific order and can be used for automating tasks. Some programs, for example, may include a batch file that runs a series of commands when the program starts up. A user can also make their own batch file to automate time-consuming tasks like moving numerous directories or renaming multiple files all at once.

Figure 10. Batch file.

Source: Image by thenounproject.com.

The COMMAND.COM program, which is included with DOS and Windows, is used to run batch files. As a result, batch files can only be used in the Windows operating system. Other scripting tools for Macintosh and Unix, such as AppleScript and Unix shell commands, can be used for comparable tasks. Because batch files contain executable commands, it is critical not to open batch files from unknown sources on your hard drive or in e-mail attachments. Because batch files contain executable commands, it is critical not to open batch files from unknown sources on your hard drive or in e-mail attachments.

File Extensions: .BAT, .CMD

Batch Process

Some computer chores can be monotonous and repetitious. If a task is indeed repetitive, a batch process can be utilized to automate most of it.

A batch process executes a series of commands in order. It can be launched by a computer's operating system as a script or batch file, or by a macro or internal scripting tool within a program. An accountant, for example, may write a script to open numerous financial programs at once, saving him the time and effort of opening each one separately. The operating system, such as Windows or Mac OS, would run this type of batch task. A Photoshop user, on the other hand, may use a batch procedure to make changes to multiple photos at once. She could, for example, record a Photoshop action that resizes and crops an image. She can batch process a folder of photographs after recording the action, which will conduct the operation on all of the images in the folder. By automating repeated activities, batch processing can save time and energy. While writing the script or recording the repetitive motions may take some time, doing it once is preferable to doing it many times.

Baud

Unlike common perception, baud is not a measurement of data transfer speed; rather, it is a measurement of the number of electrical signals sent per second. For modems, networks, serial cables, and other data transfer media, Baud is used to quantify the rate of electrical signals, or "signaling components."

Some individuals mistakenly believe that baud and bits per second are the same thing. They'll state, for example, that a 28,800-bps modem sends at 28,800 baud and behave as if they know everything. However, most modems provide multiple bits of data per baud, so while the two numbers are connected, they are rarely equal. So, the next time your friend claims that his 56K v.90 modem can transport data at 56,000 baud, you may gently correct him and explain the difference between baud and bits per second.

Bcc

"Blind Carbon Copy" is the abbreviation for "Blind Carbon Copy." You type the recipient's address in the "To:" section when sending an email to just one person. You can include addresses in the "Cc:" and "Bcc:" boxes when sending a message to several recipients. The letters "Cc" and "Bcc" stand for "Carbon Copy" and "Blind Carbon Copy," respectively.

An e-mail that is copied to one or more recipients is known as a carbon copy, or "Cc'd" message. All of the addresses the message was sent to are visible to both the principal recipient (whose address is in the "To:" field) and the Cc'd recipients. The addresses in the "Bcc:" field are not visible to the main recipient or the Bcc'd recipients when a message is blind carbon copied. Carbon copying without the main recipient's knowledge is a valuable approach to allow others see an e-mail you send. It is less time consuming than sending the original message and then forwarding it to the other recipients. When copying a message to a large group of people, it's also good etiquette to utilize Bcc. This protects the e-mail addresses from being taken and used for spamming reasons by someone on the list. If it's critical that each recipient understands who your communication is addressed to, utilize carbon copy (Cc).

Bespoke

The term "bespoke" originated in England, where it referred to custom or tailored clothing. However, in recent years, the term has been applied to information technology (IT) and has come to refer to customized services or solutions.

Bespoke software, for example, is software that has been tailored to meet a specific need. Custom accounting software for a single organization or a network monitoring tool for a specific network are examples of bespoke applications. Bespoke software is classified as vertical market software since it is produced to order for a specific purpose.

Websites are another area in which bespoke is employed in the computer business. A customized website is one that is developed from the ground up to meet the specific demands of a company or organization. This could include a unique layout, custom database integration, and other unique features requested by the client. Bespoke websites take longer to construct and are more expensive than websites made from templates since they must be individually suited to a client's needs.

Finally, the term "bespoke" can be applied to hardware. Customers may be able to customize their systems purchased from firms like Dell, HP, and Apple. For example, one individual may opt for a high-end graphics card for video production, while another may opt for a simple graphics card but add additional RAM so that her machine can run multiple programs simultaneously. Bespoke systems are a term used to describe these specialized arrangements.

Beta Software

A commercial software application normally goes through a "beta" phase before being released to the public. The software is checked for bugs, crashes, errors, inconsistencies, and other issues during this step. Though beta versions of software were once only available to developers, they are now occasionally made available for public testing, usually through the software company's website. Because beta software is free, it normally expires after a certain amount of time. If you decide to try out a beta software program, don't be surprised if it has several issues and crashes your computer frequently. It is, after all, a beta version. By looking at the program's attributes, you can identify if it's still in beta. It's a beta version if the version number starts with a "b" (e.g., Version: 1.2 b3).

Binary

Binary is a two-digit (Base-2) number system used by computers to store information and perform calculations. Because digital switches inside computers can only be turned on or off, which is represented by a 1 or 0, computers employ the binary system. Even though the binary system only has one and zero digits, the two digits can be utilized to represent any integer.

For example:

A single 0 in binary represents zero.

A single 1 represents (2^0) or 1.

10 represents (2^1) or 2.

11 represents (2^1 + 2^0) or 3.

100 represents (2^2) or 4.

101 represents (2^2 + 2^+0) or 5.

110 represents (2^2 + 2^1) or 6.

111 represents (2^2 + 2^1 + 2^0) or 7.

1000 represents (2^3) or 8, and so on.

Figure 11. Representation of binary

Source: Image by Flickr

Bing

Bing is a Microsoft-developed search engine. It evolved from "MSN Search," later renamed "Windows Live Search," and finally "Live Search."

In 2009, Microsoft decided to rebrand their search engine and give it the moniker "Bing," which is both easy to remember and pronounce.

Bing, like Google, Yahoo!, and other search engines, provides a distinct web search experience. The main page, for example, features a bespoke image or video with related information that is updated daily. You may also link your Facebook account to your Bing login, allowing Bing to show you pages that your friends enjoy in the search results.

Bing also offers a variety of specialized searches, such as photos, videos, commerce, news, and maps, in addition to online searches. Microsoft Translator, a language translation service, is also available. Bing is available for free at Bing.com.

Biometrics

Biometrics are technologies that detect and recognize physical traits in humans. Biometrics is frequently used interchangeably with "biometric authentication," a sort of security authorization based on biometric input in the IT sector.

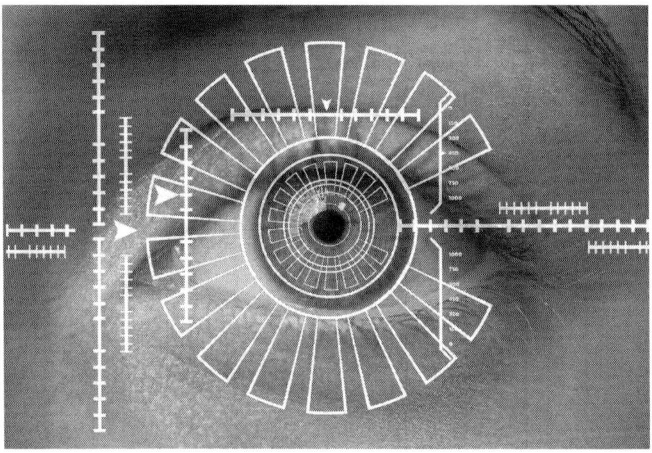

Figure 12. Biometrics.

Source: Image by Pixabay.

Biometric authentication comes in a variety of forms. Fingerprint scanning, retinal scanning, facial recognition, and speech analysis are all common instances. A facial recognition system, for example, captures an image of a person's face using a camera. Biometrics software is then used to record and process the image. The software tries to match the scanned image with a photo from a user's photo database. If the scan gets close enough to a specific user, that individual will be given permission to proceed.

A biometric scan is often used in the same way as a login. Some computers, for example, feature a fingerprint scanner that allows you to verify your identity by swiping your finger across a sensor. Instead of typing a login and password, you can authorize yourself using a finger scan. As an alternative to providing a unique pin number, several retail locations now use finger scanners to verify people's identities. In some high-security government and office buildings, retinal scans may be required to get entry to sections. In some circumstances, in addition to a biometric scan, a key card, passcode, or login is required for added security.

BIOS

"Basic Input/Output System" is the abbreviation for "Basic Input/Output System." Most individuals will never need to meddle with a computer's BIOS, but knowing what it is might be useful. The BIOS is a program that is pre-installed on Windows-based computers (but not on Macs) and is used to start up the computer. Even before the operating system is loaded, the CPU contacts the BIOS. After that, the BIOS examines all your hardware connections and locates all your devices. The BIOS then loads the operating system into the computer's memory and completes the boot-up procedure if everything is in order.

It can't survive on a hard drive since the BIOS manages them, and it can't live in RAM because it's available before the computer powers up. So, where can you find this incredible, yet mysterious BIOS? It is stored in the computer's ROM (Read-Only Memory) chip, which is an erasable programmed read-only memory. So, as soon as you turn your computer on, the CPU accesses the EPROM and gives control to the BIOS.

After the computer has booted up, the BIOS is also used. It serves as a link between the CPU and the input/output (I/O) devices. Your programs and operating system don't need to know exact data (such hardware addresses) about the I/O devices attached to your PC because of the BIOS. Only the BIOS needs to be changed when device details change. When your machine boots up, you can make these modifications by going into the BIOS. As soon as your computer starts up, hold down the DELETE or F2 key to reach the BIOS.

Bit

The computer term "bit" comes from the phrase "Binary DigIT. A bit is the smallest unit of computer data and represents a single digit number in base-2 (a zero or a one). A whole page of text contains approximately 16,000 bits.

It is important not to confuse bits with bytes. Both are used to measure amounts of data, but it takes eight bits to make one byte. The most common area where bits are used instead of bytes is in measuring bandwidth (in bits per second). Why? Probably because it makes your Internet connection sound faster than it really is.

Bitmap

Bitmaps make up the majority of the images you view on your computer. As long as you sit a decent distance away from the screen, a bitmap is a map of dots, or bits (hence the name), that appears to be an image. BMP (raw bitmap format), JPEG, GIF, PICT, PCX, and TIFF are all common bitmap file types. When you zoom in on a bitmap image, it looks to be incredibly blocky since bitmap images are made up of a collection of dots. Vector graphics (made with software like Freehand, Illustrator, or CorelDraw) may be scaled up without becoming blocky.

Bitrate

The rate at which bits are transported from one location to another is referred to as bitrate. To put it another way, it measures how much data is sent in a certain amount of time. Bitrate is typically expressed in bits per second (bps), kilobits per second (Kbps), or megabits per second (Mbps) (Mbps). A DSL connection, for example, may be capable of downloading data at 768 kbps, but a Firewire 800 connection can transmit data at 800 Mbps.

The quality of an audio or video file can also be described by bitrate. An MP3 audio file compressed at 192 Kbps, for example, will have a higher dynamic range and may sound slightly clearer than one compressed at 128 Kbps. This is because each second of playback requires additional bits to describe the audio data. A video file compressed at 3000 Kbps will also look better than a video file compressed at 1000 Kbps. The bitrate of an audio or video file is measured in the same way that the resolution of an image is measured.

BitTorrent

BitTorrent is a peer-to-peer (P2P) file sharing technology that aims to cut down on the amount of bandwidth needed to transport content. This is accomplished by spreading file transfers across numerous computers, reducing the average bandwidth consumed by each computer. If a user starts downloading a movie file, the BitTorrent system will look for other computers that have the same file and begin downloading it from all of them at the same time. Downloading from numerous computers can greatly boost the file transfer rate because most ISPs offer far quicker download rates than upload speeds.

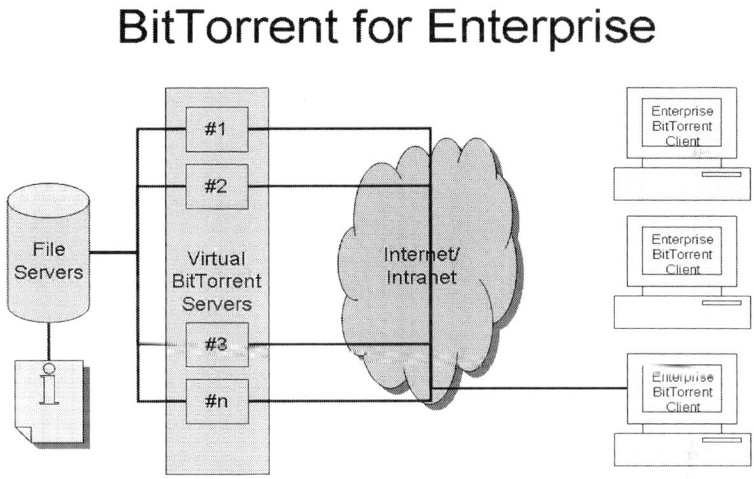

Figure 13. BitTorrent.

Source: Image by Flickr.

A BitTorrent client, which is a software program that connects to the BitTorrent network, is required to use the BitTorrent protocol. You can use the client application to look for files and start downloading torrents, which are downloads that are currently in progress. Torrents that have been paused or stopped can usually be resumed using most BitTorrent clients. When downloading large files, this is extremely useful.

Blob

The term "blob" refers to a binary large object (BLOB) that is used to store data in databases. A blob is a binary data storage data type. In contrast to most other database data types, such as integers, floating point numbers, characters, and strings, which store letters and numbers, this one store letters and numbers. Blobs can be used to store photos or other multimedia files because they can store binary data. A photo album, for example, may be saved in a database with photos recorded as blobs and descriptions stored as strings.

Blobs demand substantially more space than other data types since they are used to hold items like photos, audio files, and video clips. The amount of data a blob can hold varies by database type, however some databases allow blobs to be many gigabytes in size. Now that is a big blob!

Blog

This term refers to a list of journal entries put on a Web page and is short for "Web Log." Anyone who understands how to develop and publish a Web page can start a blog. Some Web providers have made it even simpler by providing an interface that allows users to just input a text entry and tap "publish" to publish their blog.

Figure 14. Blog.

Source: Image by Wikimedia.

Many people (often young children and adults) have found a new presence on the Internet as a result of the ease with which they can create a blog. Individuals can now share their personal sentiments and experiences with thousands of people around the world instead of writing confidential entries in a book that no intends to see. Blogs are often updated on a daily, monthly, or irregular basis. "Blog" may also be used as a verb, as in "Wow, Matt blogged a lot last week."

Blu-ray

Blu-ray, such as CDs and DVDs, is an optical disc format. It was designed for storing enormous amounts of data and recording and playing back high-definition (HD) video. A single Blu-ray disc can carry up to 25 GB of data, compared to 700 MB on a CD and 4.7 GB on a standard DVD. Even a rare double-sided, dual-layer DVD can only carry 17 GB of data. 50 GB of data will be stored on dual-layer Blu-ray discs. That's the equivalent of 4 hours of high-definition television.

Figure 15. Blu-Ray.

Source: Image by Flickr.

Because of the blue lasers used in the drives, Blu-ray discs can carry more data than regular optical media. Although the laser is blue-violet, "Blu-ray" is a little easier to say than "Blu-violet-ray." The wavelength of the blue-violet laser is shorter than that of the red lasers used in CDs and DVDs (405 nm compared to 650 nm). This allows the laser to focus on a smaller region, allowing far more data to be packed onto a disc the same size as a CD or DVD. As more people switch to HDTV, supporters of the Blu-ray technology predict that Blu-ray players will replace VCRs and DVD recorders. Visit Blu-ray.com for the most up-to-date Blu-ray news.

Bluetooth

This wireless technology allows Bluetooth-enabled devices to communicate with one another. It connects desktop and laptop computers, PDAs (such the Palm Pilot or Handspring Visor), digital cameras, scanners, cellular phones, and printers over short distances.

Infrared used to serve the same purpose as Bluetooth, but it had a few disadvantages. If an object were to be placed between the two communication devices, for example, the transmission would be disrupted. (If you've ever used a television remote control, you've probably observed this limitation.) Infrared transmission was very slow, and devices were frequently incompatible with one another.

Figure 16. Symbol of Bluetooth.

Source: Image by Wikimedia.

All of these limitations are taken care of via Bluetooth. Because the system is based on radio waves, items or even walls can be placed between the communication devices without disrupting the connection. Bluetooth also employs a common 2.4 GHz frequency, ensuring that all Bluetooth-enabled devices are interoperable. The sole disadvantage of Bluetooth is that its range is restricted to 30 feet due to its high frequency. While this is fine for transferring data within the same room, you may be out of luck if you are walking in your backyard and want to transfer your address book from your cell phone to your basement computer. The low range, on the other hand, can be viewed as a positive feature because it increases the security of Bluetooth connection.

BMP

"Bitmap" is abbreviated as "Bitmap." "Bump," "B-M-P," or simply "bitmap image" are all possible pronunciations. The BMP format is a raster graphic format for saving image files that is widely used. It was first introduced on the Windows platform, but many programs now recognize it on both Macs and PCs.

Without any compression, the BMP format stores color data for each pixel in the image. A 10x10 pixel BMP image, for example, will have color data for 100 pixels. This way of storing image information yields enormous file sizes yet allows for crisp, high-quality visuals. The JPEG and GIF formats are both bitmaps, but they use image compression technologies to reduce file sizes dramatically. As a result, JPEG and GIF images are commonly utilized on the Web, while BMP images are frequently used for printing.

Bookmark

An Internet bookmark functions similarly to a physical bookmark in that it serves as a marking for a website. (They're called "Favorites" in Internet Explorer.) To go to a certain site in a Web browser, simply select a bookmark from the browser's Bookmarks menu. You won't have to repeat the process of putting in the Internet address every time you visit one of your favorite sites this way. Also, who remembers those 200-character addresses anyway?

Figure 17. Bookmark symbol.

Source: Image by Thenounproject.com.

When you're on a website you want to bookmark, most browsers let you make a bookmark by selecting "Add Bookmark" from the Bookmarks menu. Wow, that's a lot of "bookmarks" in one line. That's what happens when nouns and verbs are used interchangeably.

Boolean

This is the logic that computers use to determine if a statement is true or false. There are four main Boolean operators: AND, NOT, OR, and XOR. Below are some examples of how the 4 operators work:

x AND y returns True if both x and y are true, otherwise the expression returns False.

NOT x returns True if x is false (or null) and False if x is true.

x OR y returns True if either x or y or both are true; only if they are both false will it return False.

x XOR y returns True if either x or y are true, but not both. If x and y are both true or false, the statement will return False.

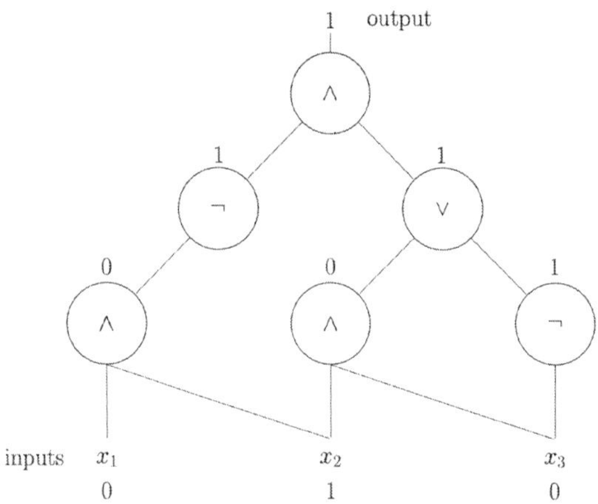

Figure 18. Boolean.

Source: Image by Wikipedia.

While Boolean expressions are used to power computers' CPUs, they can also be used by users. Many search engines, for example, accept Boolean operators in search queries when looking for information on the Internet (i.e., "Yamaha AND piano NOT motorcycle"). In software development, Boolean expressions are frequently used to control loops and variables.

Boot

You kick a football a long far when you boot it. You just switch on a computer when you boot it. It's not a good idea to kick your computer too hard, even if you're inclined to do so. The word "boot" comes from the term "bootstraps," which was sometimes used to put on one's footwear. Similarly, "booting" a computer brings it up to speed.

To put it simply, booting a computer is turning it on. The "boot process" begins when the computer's power is turned on. The start-up instructions are loaded from the computer's ROM, and then the operating system is loaded from the current boot disc. The boot disc is commonly a hard drive inside the computer, although it can also be an external drive, a CD or DVD-ROM, or even a floppy disc. Once the operating system software is loaded, the boot process is complete and the computer is ready to be used.

Boot Disk

A boot disc isn't a computer disc that looks like a boot. Most disc drives would have a hard time reading it if it was. A boot disc, on the other hand, is a disc that a computer can use to start up or "boot" from. An internal hard drive is the most popular sort of boot disc, and it's where most computers start up from. During the boot process, the operating system installed on the hard drive is loaded.

Most computers, on the other hand, allow you to boot from external Firewire hard drives, CD-ROMs, DVD-ROMs, and floppy discs. These discs must have an operating system installed that the machine can understand to serve as boot discs. This can be a full-fledged operating system such as Windows or Mac OS X, or a small utility operating system like Norton Utilities or Disk Warrior.

When the operating system on the internal hard drive fails to load, CD and DVD boot discs are frequently used to restart the machine. When faulty data blocks or other mistakes occur on the disc, this can happen. You may typically repair the hard drive and restart it using the whole operating system by utilizing a disc repair program from a CD or DVD.

Boot Sector

The first section of a hard disc or other data storage medium is known as the boot sector. It houses the master boot record (MBR), which the computer consults throughout the boot process. A partition map, which defines each disc partition, may also be included in the boot sector. For example: "The computer could not start up from the hard disk because the boot sector was damaged."

Boot Sequence

When a computer first turns on, it goes through a variety of procedures. A "boot sequence" is the name given for this series of events. The computer activates the necessary hardware components and loads the needed software during the boot procedure so that a user can actively interact with the system.

The boot sequence starts by accessing the computer's BIOS on Windows PCs or the system ROM on a Macintosh. Basic instructions teach the computer how to boot up and are stored in the BIOS and ROM. These instructions are subsequently sent to the CPU of the computer, which begins loading data into the system RAM. The computer begins loading the operating system into system memory once a valid boot disc or start-up disc is detected. The computer is ready to use once the operating system has finished loading.

Depending on the computer's configuration, the boot sequence might take anything from a few seconds to several minutes. The boot time may be much greater if the computer is booted from a CD or DVD than if it is booted from a hard drive. Additionally, if your computer was shut off unexpectedly, the start-up time may increase while the system does additional checks to ensure everything is in working order.

Bot

This is an automated software program that can execute certain commands when it receives a specific input (like a ro-"bot"). Bots are most often seen at work in the Internet-related areas of online chat and Web searching. The online chat bots do things like greet people when they enter a chat room, advertise Web sites, and kick people out of chat rooms when they violate the chat room rules. Web searching bots, also known as spiders and crawlers, search the Web and retrieve millions of HTML documents, then record the information and links found on the pages. From there, they generate electronic catalogs of the sites that have been "spidered." These catalogs make up the index of sites that are used for search engine results.

Figure 19. Bot.

Source: Image by Pixabay.

Botnet

A botnet is a collection of computers controlled by a single computer. While some botnets are built for lawful cluster computing reasons, the vast majority of botnets are designed for malevolent objectives. Sending spam messages, propagating malware, and attacking Internet servers are just a few examples. For example: "When he ran the antivirus software, he realized his computer had been part of a botnet."

Botnets are created when hackers compromise the security of numerous computers and install bots (automated programs) on each one. This is frequently performed by taking advantage of a security flaw in the operating system or a piece of software. Because botnet activity is often hidden from the user, most people are unaware that their computers have been compromised. As a result, it's a good idea to turn on your system firewall and install antivirus software that monitors your computer for strange activities on a frequent basis.

Figure 20. Botnet.

Source: Image by Thenounproject.com.

Bounce

The term "bounce" has several different IT related meanings, yet none of them include bouncy balls. The most common definition of bounce used in the computer world refers to e-mail messages.

1. Returning E-mail

When you send an e-mail message to someone else, the message is processed by the mail server and delivered to the appropriate user's mailbox. If you send an email to "mrman@mail.com," for example, the mail.com server looks for a user named "merman" to deliver it to. If the user does not exist, the message may be bounced back to the sender with the message "Sorry, that user does not exist." These messages often come from "Mail Delivery Subsystem" and have a subject line that reads "Returned mail: see transcript for details."

If you get a bounced message, double-check the e-mail address you used to send the message and make sure it's correct. If the address is valid, reading the body of the bounced message for extra information may be helpful. "User quota over limit," for example, indicates that the recipient has exceeded his or her e-mail quota and must delete some messages and/or attachments in order to receive new mail. If this is the case, you may want to phone the person or use a different e-mail address to inform him or her that he or she needs to perform some Inbox maintenance.

2. Restarting a Computer

The process of rebooting or restarting a computer is also known as "bounce." After installing new software, for example, a workstation may need to be bounced. A Web server may also be bounced if the websites hosted on it are not responding properly.

3. Exporting Audio

The term "bounce" can also refer to the process of converting many tracks in an audio mix to one mono or two stereo recordings. This aids in the consolidation of audio tracks that have been mixed. Bouncing audio tracks reduces the amount of processing power required because the computer only has to process one track instead of all of them. Digital Performer is the primary audio software program that uses bouncing to export audio.

4. Hiding a Network Connection

Finally, in networking, the term "bouncing" can refer to a technique for concealing the source of a user's network connection. BNC is a common abbreviation for this form of bouncing. A "bouncer" is someone who bounces his network connection, but this is not the same person who verifies your ID at the bar.

BPS

"Bits Per Second" is the abbreviation for "Bits Per Second." (The "b" is capitalized because it refers to bits rather than bytes.) The conventional way of measuring how fast data passes via a network or phone system is in bits per second. A 56K modem, for example, can theoretically transmit data at 56,700 bits per second.

Bridge

A bridge is erected to connect two land masses when a road needs to extend across a river or valley. Because the average car can't swim or fly, the bridge allows cars to continue driving from one land mass to the next.

A bridge serves a similar job in computer networking. It joins together two or more local area networks (LANs). The bridge is used by the cars, or data in this example, to move to and from different parts of the network. Although the device resembles a router, it does not evaluate the data being sent. As a result, bridges are usually quick in transferring data, but they aren't as adaptable as routers. For example, unlike most routers, a bridge cannot be utilized as a firewall. A bridge operates at the "data link layer," or level 2 of the OSI (Open Systems Interconnection) networking reference model, and can transport data across different protocols (for example, a Token Ring and Ethernet network).

Broadband

This is a type of high-speed data transmission in which a single cable may simultaneously convey a significant volume of data. Cable modems (which use the same connection as cable TV) and DSL modems are the most frequent types of Internet broadband connections (which use your existing phone line). Broadband has begun to supplant baseband, the single-channel technology that was originally utilized in most computer networks, due to its multiple channel capability. So now you'll know what corporations like AT&T and SBC are talking about when they put those fancy "broadband" advertising in your face.

Figure 21. Broadband.

Source: Image by Thenounproject.com.

Buffer

A buffer is a storage area for data that is only kept for a brief period, usually in the computer's memory (RAM). A buffer's job is to keep data until it's time to use it. When you download an audio or video file from the Internet, for example, it may put the first 20% of it into a buffer before starting to play. While the video is playing, the computer is downloading the rest of the video and saving it in the buffer. There is less of a possibility that the audio or video may stall or skip when there is network congestion because the clip is being played from the buffer rather than directly from the Internet.

Buffering is also utilized to increase computer speed in other areas. A buffer is used by most hard drives to provide for faster access to the data on the disc. Before being displayed on the screen, video cards transmit images to a buffer (known as a screen buffer). Buffers are used by computer programs to store data while they are running. Computers would work far less efficiently without buffers, and humans would have to wait a lot longer.

Figure 22. Buffering symbol.

Source: Image by Thenounproject.com.

Bug

A bug is a software program error in the computer world. It could cause a software to quit unexpectedly or behave in an unexpected way. A simple problem, for example, could cause a button in a program's interface to not respond when you click it. Due to an infinite calculation or memory leak, a more serious problem may cause the program to hang or crash.

Bugs in a program's source code can be syntax or logic problems from the developer's perspective. A debugger, as the name implies, is a development tool that can be used to rectify these mistakes. The problems will be detected by the user if errors are not discovered before the software is compiled into the final application.

Because bugs can degrade a program's usefulness, most programs undergo extensive testing before being released to the public. Commercial software, for example, frequently goes through a beta phase in which several users extensively test all components of the program to ensure that it runs properly. Once the program is determined to be stable and free from errors, it is released to the public.

In fact, even after being properly tested, most programs are not completely error-free, as is common knowledge. As a result, software developers frequently produce "point updates" (e.g., version 1.0.1), which contain bug patches for issues discovered after the software was released. Programs that are especially "buggy" may require multiple point updates (1.0.2, 1.0.3, etc.) to get rid of all the bugs.

Burn

When you "burn a disc," you perform the action of writing data on it. If you were taking the SAT test, the parallel would go as follows:

Hard Disk::: write

Burn a CD/DVD

Because the CD-writer, or burner, literally burns the data onto a writable CD, the term "burn" is employed. A CD-laser writer can be turned up to a higher power level than a standard CD-ROM laser. It can engrave hundreds of 1s and 0s in the form of binary language onto a CD because of this property.

As a result, people refer to "burning" tunes or files to CDs. They could just say they're "writing" the data to a CD and get away with it, but people seem to prefer "burning."

Bus

While a bus's wheels run "round and round," the data on a computer's bus goes up and down. Each bus in a computer is made up of a series of wires that allow data to flow back and forth. Most computers have multiple buses that carry data to various portions of the machine. Each bus has a size, specified in bits (such as 32-bit or 64-bit), that controls how much data can possibly be sent across it at once. Buses also have a speed, measured in megahertz, that governs how quickly data can be transmitted across a system.

The frontside bus is the computer's primary bus, and it connects the CPU to the rest of the motherboard's components. Data can be moved to and from expansion cards, like as video cards and other I/O devices, using expansion buses like PCI and AGP. While a computer has numerous buses, the frontside bus is the most significant since it affects how quickly data can enter and exit the processor.

Byte

A byte is an 8-bit representation of a single character in computer memory. This term should not be confused with "bite," as in biting into a cookie, because they are completely distinct. Bytes are used to measure file sizes, hard drive capacity, and computer memory, while bits are used to measure data transfer speeds. Megabytes, gigabytes, and terabytes are units of measurement for larger amounts of data. One kilobyte, for example, is equivalent to 1,024 bytes.

C

C/C++

C is a mid-1970s high-level programming language. It was created to write Unix programs, but it is now used to create apps for almost every platform. C is easier to comprehend, more flexible (it may be used for a wide range of purposes), and memory efficient than most preceding languages. C++ is a programming language based on the C language. It is pronounced "C plus plus." C++ has a syntax that is essentially comparable to C, but it has object-oriented capabilities that allow it to be used with other languages.

Cable Modem

A cable modem is substantially faster than a traditional dial-up modem for connecting to the Internet. Cable modems can enable data transfer rates of up to 30 Mbps, but 56K modems can only accept data at roughly 53 Kbps. That's a 500-fold increase in speed. To save bandwidth, most ISPs limit their subscribers' transfer rates to less than 6 Mbps.

Figure 23. Cable Modem.

Source: Image by Pxhere.com.

A cable modem differs from a dial-up modem in another pertinent way: it does not connect to a phone line. The cable modem, as the name implies, connects to a local cable television line. This allows cable modems to maintain a constant Internet connection. As a result, you will not have to dial your ISP every time.

Cable modems are normally separate devices with a much more complicated design than dial-up modems, but certain variants can be integrated within a computer. Instead of connecting to a serial port like a dial-up modem, cable modems connect to a normal Ethernet port to send data at the highest feasible speed.

Cache

This term is pronounced like "cash" rather than "catch" or "cashé." A cache saves recently used data in a location where it may be accessed quickly. A cache, for example, is used by a Web browser like Internet Explorer to save recently seen pages, images, and URLs of recently visited Web sites on your hard drive.

When you visit a page, you have previously visited, the pages and photos do not have to be downloaded all over again thanks to this clever method. Caching Web Sites can greatly speed up Web browsing because accessing your computer's hard disc is much faster than accessing the Internet. The amount of the cache may usually be adjusted in the browser options of most Web browsers.

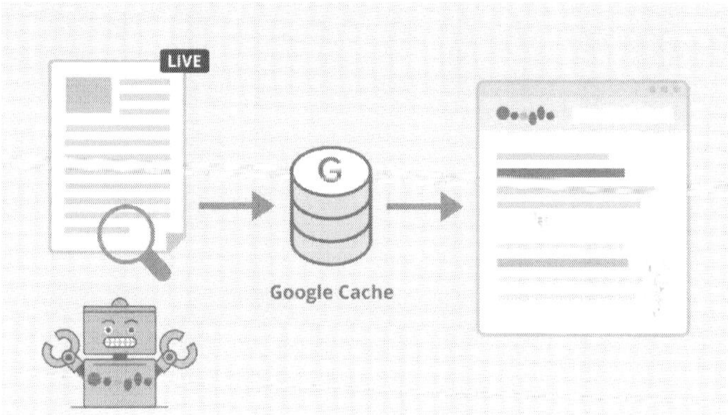

Figure 24. Cache.

Source: Image by Seoability.net.

A disc cache is another typical form of cache. This keeps information you've recently read from your hard disc in the RAM, or memory, of the computer. Because reading data from the hard disc is slower than accessing RAM, this can let you access common files and directories on your hard drive much faster.

A processor cache is another sort of cache that stores small bits of data immediately adjacent to the processor. This aids in the efficient processing of common instructions, resulting in a reduction in computing time.

CAD

Stands for "Computer-Aided Design" Engineers and architects see it as the greatest invention of all time. CAD software is now utilized for almost all three-dimensional design work. Designers can, in a much more rapid and accurate manner, convert an object into an electronic representation than they can using a pencil and paper. Even better, items generated with CAD software may be quickly moved, scaled, and rotated with a pencil and paper, you can only erase so much before it starts to smudge.

Figure 25. Computer-aided design.

Source: Image by Picpedia.

Camera RAW

The images you take with most digital cameras are processed and compressed right after they are taken. This is advantageous since it reduces file sizes (via JPEG compression) and handles color adjustment, such as white-balance, tint, and exposure, so you don't have to. Professional photographers, on the other hand, prefer more control over how each shot is treated. As a result, many high-end cameras are capable of shooting in RAW mode. This option does not compress the images at all and does not manipulate them at all.

Camera RAW files take up more space than JPEG photographs since they are uncompressed. It is similar to developing a film negative in a dark room. The negative is the RAW file, and the dark chamber is the computer. You have complete control over temperature adjustments (for white-balance), tint, color, and exposure using a RAW file. Most image-viewing tools, however, will not open Camera RAW files since they are not normal photos. As a result, most high-end camera manufacturers provide Camera RAW editing software with their products. These apps let you open RAW files, perform necessary processing, and save the results in standard picture formats including bitmap, TIFF, and JPEG.

While shooting in Camera RAW mode gives you a lot of control over your photographs, each photo you snap requires additional software and time to access and edit. Using the camera's built-in processing is the best solution for most casual photography.

Caps Lock

Caps lock is a function found on nearly all keyboards that capitalizes each letter written on the keyboard when activated. The caps lock key, unlike most other keys on the keyboard, is a "toggle key," meaning it is always in one of two states: on or off. When the switch is turned off, the letters written are entered as usual. When the feature is enabled, each letter written is sent as an upper-case character. The input from the rest of the keys, such as numbers and symbols, is unaffected by caps lock.

It is crucial to know if caps lock is on or off because it alters the input of the characters on the keyboard. As a result, most keyboards have a light on the caps lock key that illuminates when the key is pressed, signaling that caps lock is activated. Because passwords are frequently case-sensitive, it's a good idea to avoid using caps lock when entering login information. When typing e-mails, it's also a good idea to avoid using caps lock because it makes it appear as if you're yelling.

Captcha

"Captchas may be used on online forms to prevent automated programs from inputting data." A captcha is a challenge-response test that is used to detect whether a user is human or a bot. A standard captcha consists of a distorted text image and a form field in which the user must enter the text. Captchas are often seen at the bottom of online forms, and they must be completed before the form can be submitted. Webmasters can prevent automated programs from delivering spam or other harmful material through online forms by requiring users to decipher and enter the captcha text.

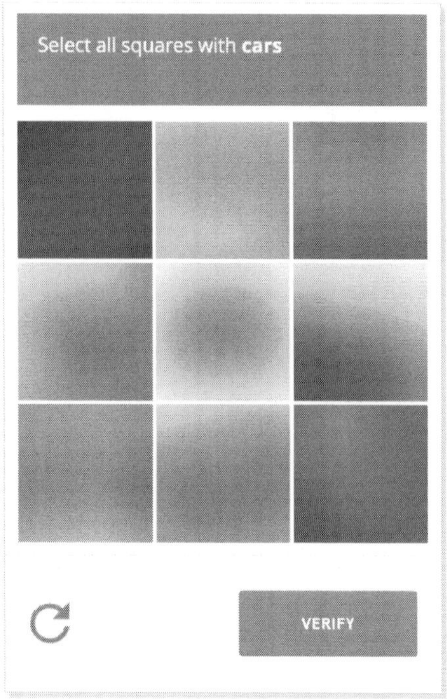

Figure 26. Captcha is used for security purpose.

Source: Image by Wikimedia.

Card Reader

The phrase "card reader" refers to any input device that reads flash memory cards. It can be a standalone USB device or integrated into a computer, printer, or multifunction device. Most multifunction printers, scanners, and copiers now have a card reader.

Figure 27. Card reader.

Source: Image by Pixabay.

Compact flash (CF), secure digital (SD), and Sony's Memory Stick are among the memory card formats that most card readers accept. Other card formats accepted by some card readers include XD, SmartMedia, Microdrive, and Memory Stick Pro Duo.

A card reader, as the name implies, is used to read data from a memory card. When you insert a memory card into a card reader, it frequently appears as a mounted disk on your computer. The contents of the memory card can then be viewed by double-clicking the card's icon. On Macintosh computers, this symbol appears on the desktop, while on Windows systems, it appears in "My Computer."

When you insert a memory card into your card reader, a photo organizing program may automatically launch because memory cards typically include photographs from digital cameras. This allows you to quickly import your photos into your photo album. If you don't want to use the program to import images, simply close it and the card will remain mounted on your computer.

Make sure you unmount or "eject" the disk before physically removing the card once you've decided to remove it. This will help keep the card's data from becoming easily corrupted.

Cc

Stands for "Carbon Copy." The phrase derives from carbon copying, which is the process of transferring writing from one piece of paper to another using a piece of carbon paper (often used when filling out forms). However, e-mail is now widely referred to as "e-mail." In the "To:" area of an e-mail message, you usually type the recipient's address. You can use the "Cc:" box to add other addresses to the message if you wish to send it to one or more other people. The e-mail will be sent to the address in the "To:" area as well as each of the addresses given in the "Cc:" field.

The "Cc:" option is often used in business communications when a message is intended for one person, but is relevant to other people as well. A shop employee, for example, may send an e-mail to another employee stating that he is available to work for her on a specific day. In the "Cc:" field, he might enter his manager's and assistant manager's e-mail addresses to inform them that he will be working the shift. Similarly, a team member working on a product design may e-mail his supervisor with the most recent design revisions and "Cc:" the rest of his team to let them know the e-mail was received.

"CCing" (yep, it's a verb) is a simple way to include other persons in your e-mail communications. It saves time because you don't have to send individual messages to each address. However, keep in mind that when you Cc an email, all recipients will be able to view the other addresses to which the letter was sent. Instead, utilize Blind Carbon Copy (Bcc) to disguise the additional addresses.

CCD

Stands for "Charged Coupled Device." CCDs are sensors that record still and moving images in digital cameras and video cameras. The CCD captures light and turns it into digital data, which the camera records. As a result, a CCD is frequently referred to as the digital equivalent of film.

The resolution of a CCD sensor determines the quality of an image taken by it. The resolution of digital cameras is measured in Megapixels (thousands of pixels). As a result, a digital camera with 8 megapixels may collect twice as much data as one with 4 megapixels. As a result, the image is larger and more detailed.

CCDs in video cameras are typically sized by their physical dimensions. Most consumer digital cameras, for example, have a CCD that is about 1/6 or 1/5 of an inch in size. CCDs in more expensive cameras may be 1/3 of an inch or greater. The larger the sensor, the more light it can collect, resulting in superior footage in low-light conditions.

CCDs for capturing red, green, and blue hues.

CD

"CD" is the abbreviation for "Compact Disc." CDs are circular discs with a diameter of 4.75 inches (12 cm). Sony and Philips proposed the CD standard in 1980, and the technology was released in the United States in 1983. CDs have a capacity of up to 700 megabytes of data or 80 minutes of audio. The data on a CD is stored on the disc as microscopic notches that are read by a laser from an optical drive. The drives convert the notches (which indicate 1s and 0s) into data that may be used.

Audio CDs were the first CDs, and they gradually supplanted audio tapes (which earlier replaced records). Audio CDs have the advantage of allowing the user to navigate around the disc at will. CDs can also be listened to an unlimited number of times without losing quality. Audio tapes can start to lose quality after listening to them as few as ten times. This is because the laser that reads the data on a CD doesn't put pressure on the disc, whereas the play heads on a tape deck slowly wear away the magnetic strip on the tape.

CD-ROMs first appeared on the computer scene in 1985. CDs quickly became the most popular software format because they could store significantly more data than floppy discs (700 MB vs. 1.4 MB). CD-R (CD-Recordable) technology was introduced in 1988, allowing computer users to create their own CDs. However, it was not until the late 1990s that this technique became widely used. A 3" CD known as "CD-3" is also available, which can be read by most tray-loading CD-ROM machines.

CDR

Stands for "Compact Disc Recordable" CD-R discs are blank CDs that can be used to record data that has been written to them by a CD burner. Because CD-Rs are frequently used to record audio that can be played back by most CD players, the term "recordable" is employed. However, because CD-RWs can be erased, they don't store data reliably for as long as CD-Rs do. Therefore, you should use regular CD-Rs for long-term backups.

Figure 28. Compact disc recordable.

Source: Image by Pixabay.

CDMA

Stands for "Code Division Multiple Access." CDMA is a wireless transmission method invented by the English allies during World War II to circumvent transmission jamming. Qualcomm patented the concept and commercialized it as a digital cellular technology after the conflict ended. Many cell phone providers now employ CDMA as a prominent mode of communication.

CDMA, unlike GSM and TDMA, communicates across the entire frequency range accessible to it. It does not allocate each user on the communications network a distinct frequency. During World War II, this method, known as multiplexing, made it difficult to jam signals. There is greater bandwidth available since CDMA does not limit each user's frequency range. More users can communicate on the same network at the same time than if each user had their own frequency band.

Because CDMA is a digital technology, analog audio signals must be converted to digital before being broadcast over the network. CDMA is a wireless communication standard that is used in 2G and 3G networks and runs at frequencies ranging from 800 MHz to 1.9 GHz.

Cell

In a spreadsheet, a cell is defined as the intersection of a row and a column. Cell C7, for example, is the cell in the seventh row of column C. Cells can hold a variety of data kinds, including numbers, dates, and text. Data that is directly entered by the user is stored in static cells. Formulas are stored in dynamic cells, which show the results of computations dependent on the contents of other cells in the spreadsheet. For example: "The annual revenue is located in cell D21 of the company's financial spreadsheet."

Certificate

An SSL certificate, also known as a secure certificate, is a file that identifies a website and is installed on a secure Web server. This digital certificate verifies the company's or merchant's identity and legitimacy, giving online customers confidence that the website is safe and secure.

A web server that is used to identify a website. This digital certificate verifies the company's or merchant's identity and legitimacy, giving online customers confidence that the website is safe and secure. The companies and their websites are verified by a third party, such as Verisign or Thawte, to ensure that they are legitimate (that they are who they say they are).

Figure 29. SSL certificate.

Source: Image by Flickr.

The verification provider will issue an SSL certificate (for a minor charge of a few hundred dollars) once the authenticity of an organization and its connected website has been established. When a user enters a secure part of the website, this digital certificate is put on the Web server and visible. When the URL begins with "https," you know you're on a secure page. To see the certificate, hover your mouse over the lock icon at one of the browser window's boundaries.

Digital certificates do not last indefinitely because they certify a company's current state. SSL certificates usually have a one-to-three-year expiration period. If the certificate is not renewed in a timely manner, a warning box stating "This website's certificate has expired" may appear.

This error has nothing to do with you or your computer; it appears because the SSL certificate for the Web server to which you connected has not been renewed. While this does not necessarily indicate that the site is bogus, it does indicate that it is not professionally run.

CGI

The term "Common Gateway Interface" refers to a collection of rules that determine how scripts and programs can be run on a particular Web server. The "Common" in "Common Gateway Interface" alludes to the fact that CGI scripts may execute on a variety of server platforms and in a variety of languages. Example – "Formmail.pl is a typical CGI script that allows users to send e-mail through a form on a website." In the root directory of each website, most Web servers contain a folder called "cgi-bin." The CGI rules and permissions must be followed by any scripts installed in this folder. A script in the cgi-bin folder, for example, may have access to system resources that other scripts do not.

Although CGI scripts are still used on many websites, Web developers are increasingly writing scripts in languages such as PHP, ASP, and JSP. Instead of being referenced from another URL, these scripts can be inserted directly within an HTML page. CGI also stands for "Computer Generated Imagery," or 3D computer graphics used in film, television, and other forms of multimedia.

Character

Any letter, number, space, punctuation mark, or symbol that may be entered on a computer is referred to as a character. For example, the word "computer" has eight characters. The phrase "Hello there." is nine characters long. Because each character takes up one byte of space, the word "computer" takes up eight bytes. The list of characters that can be typed is defined by the ASCII and extended ASCII set. Some of the symbols accessible are unusual, prompting you to exclaim, "That's quite a character!"

"Since the checksums of the CDs matched, he knew the discs were identical," says the checksum example.

A checksum is a computed number that is used to verify a file's integrity. Many techniques, such as the cyclic redundancy check (CRC) or a cryptographic hash function, can be used to construct it. When the checksums of two sets of data match, the data sets are considered identical.

While the phrase "chip" is vague, it should not be confused with the term "card." A graphics processor may be incorporated in the motherboard of a laptop, whereas a graphics card attached to a PCI or AGP slot may be found in a desktop computer. A chip can be found on a graphics card, but a card cannot be found on a chip. A CPU, similarly, may have a chip (the processor) as well as various other components. As a result, the term "chip" can be used to refer to individual components, but it should not be used to describe a group of components.

Chipset

An integrated circuit's architecture is described by a chipset. This covers the circuitry's layout, the components utilized in the circuit, and the circuit board's functionality. The chipset of a modem card, for example, is very different from the chipset of a computer's CPU.

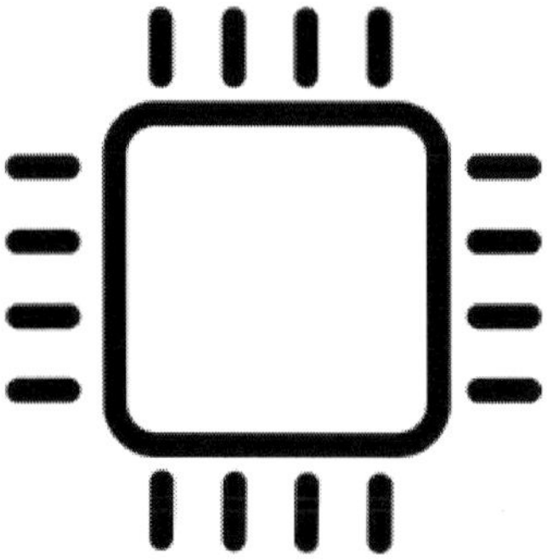

Figure 30. Chipset.

Source: Image by Thenounproject.com.

Chipsets are also variable between processors. A PowerPC processor, for example, with RISC architecture, will have a completely different chipset than an Intel CPU with CISC technology. Even though they are both built by Intel, a Pentium II and a Pentium III have somewhat different chipsets. The crucial point is that a circuit functions with other components inside the computer, regardless of the chipset it utilizes. As a result, if you replace your computer's hardware, be sure you purchase components that are made for and compatible with your machine's chipsets.

CISC

IT stands for "Complex Instruction Set Computing." This is a design for a microprocessor. The CISC architecture includes a vast number of computer instructions that range from basic to advanced and specialized. Though the design was meant to compute complicated instructions as quickly as possible, it was eventually discovered that numerous little, short instructions could do so more quickly. Reduced Instruction Set Computing (RISC) became the other major type of microprocessor architecture because of this. The Intel Pentium CPUs are mostly CISC-based with some RISC features, whereas the PowerPC processors are entirely RISC-based.

Clean Install

A clean install is an operating system (OS) installation that wipes the hard disk clean of all previous data. A clean install, unlike a standard OS upgrade, removes the existing operating system and user files throughout the installation process. When a clean install is completed, the hard disk just contains the new operating system, much like when a computer is first used. When upgrading your operating system, a clean install is rarely essential. A normal "upgrade and install," which merely upgrades the necessary files while leaving the user data in place, is significantly easier and safer. However, in rare cases, an OS upgrade is not possible due to the loss or corruption of crucial files. A clean install may be the only choice in this instance. Some users may opt to perform a clean install to ensure that no issues from the prior operating system will influence the newly installed operating system. Additionally, when installing an OS on a new hard drive or transferring ownership of a machine to another person, a clean install may be necessary.

When upgrading your operating system, both Windows and Mac OS X allow you to conduct a clean install. Near the beginning of the installation process, the installer will give you the option of a normal upgrade (usually the default option) or a clean installation. If you choose a clean install, Windows 7 also allows you to format and partition your installation disk. Before performing a clean install in Mac OS X, you can use the Disk Utility program to format or partition your drive.

Important: Because a clean install wipes out all data on the primary hard disk, it's critical to make a backup before proceeding. While it's always a good idea to have a recent backup of your data before upgrading your operating system, it's especially critical when doing a clean install.

While it's always a good idea to have a recent backup of your data before upgrading your operating system, it's especially critical when doing a clean install. It's a good idea to back up your data to an external hard drive or another computer system. It's also a good idea to double-check your backup to make sure it contains all the files you require, so you don't lose anything.

Clean room

A clean room is a highly purified and regulated room for producing equipment that is sensitive to contamination.

Microprocessors, for example, are manufactured in a clean environment because even little amounts of dust can cause manufacturing errors. As a result, in a clean space, the air is constantly filtered. Temperature, humidity, and pressure are all variables that may be controlled.

According to U.S. federal norms, clean rooms are classified into classes. A cubic foot of air in a Class 1,000 clean room contains no more than 1,000 particles larger than 0.5 microns. There are no more than 100 particles in a

Class 100 clean room. People who work in clean rooms must wear protective garments to avoid contaminating the environment. They resemble astronaut suits in appearance, but are lighter and do not require the enormous boots required for walking on the moon.

Client

Businesses have clients in the real world. Clients and servers exist in the computer world. In both local and wide area networks, the "client-server" design is prevalent. If a corporation has a server that stores the company's database, the other computers in the workplace that may access the database are called "clients" of the server. On a broader scale, your computer acts as the client that connects to the mail server when you access your e-mail from an Internet mail server. The software that functions as an interface between the client computer and the server is referred to as "client software." If you check your e-mail with Microsoft Outlook, for example, Outlook is your "e-mail client program," which allows you to send and receive messages from the server. Isn't it fascinating how everything works?

Clip Art

A collection of photographs or images that may be imported into a document or another application is referred to as clip art. Raster or vector graphics can be used to create the images. The number of photos in a clip art gallery might range from a few to hundreds of thousands.

People, objects, nature, and other categories are commonly used to classify clip art, which is especially useful when searching through thousands of photos. Most clip art images are also labelled with keywords. For example, a picture of a female teacher in a classroom may have the keywords "school," "teacher," "woman," "classroom," and "students" associated with it. You can search for images using these keywords in most clip art systems.

You can copy a clip art image to your computer's clipboard and paste it into another software, such as Photoshop or Microsoft Word, once you've found one you want to use. You might be able to save the image to your Desktop or another folder on your computer. The majority of clip art is royalty free, which means you can use it without paying royalties to the artists who created it. So, if you spend $50 for a clip art package with 50,000 images, you are only paying a tenth of a penny each image. That's a fantastic deal.

Clipboard

The clipboard is a memory area on your computer where copied data is stored. This could be a word selection, an image, a file, or any other sort of data. When you use the "Copy" command, which is usually found in the Edit menu of most programs, it is copied to the clipboard. The "Paste" command, which is also found in most programs' Edit menu, can be used to paste data which has been copied to the clipboard into a document or program. An image copied to the clipboard from your photo album, for example, might be put into a photo editing tool like Photoshop. A Web address can be copied from an e-mail to the clipboard and pasted into the address field of your Web browser.

You can examine what data is stored in the clipboard in several programs. For example, you can choose "Show Clipboard" from the Edit menu in Mac OS X's Finder. When you copy data to the clipboard, any existing data in the clipboard is usually overwritten by the new data. Because the clipboard data is kept in RAM, when your computer is shut down or restarted, it is also destroyed.

CLOB

CLOB stands for "Character Massive Object," and it refers to a data structure for storing large volumes of character data. Several current database management systems (DBMS) support CLOBs, albeit some SQL functions cannot be performed on CLOB columns. The CLOB data type is similar to the BLOB data type, but it stores text instead of binary data.

"Most CPU processes take many clock cycles," says the example. A clock cycle, often known as a clock tick, is a unit of processor activity that can be completed during one increment of the CPU clock. Load, store, and jump procedures are all common clock cycle actions. To complete a calculation, such as adding or subtracting two integers, these actions are usually conducted in a sequence.

A processor's clock speed is expressed in clock cycles per second, or "Hz." For example, a CPU that completes three billion clock cycles per second has a clock speed of 3,000 megahertz, or 3 gigahertz.

Clock Speed

The clock speed of a processor is the rate at which a processing cycle can be completed. Megahertz and gigahertz are the most common units of measurement. One million cycles per second equals one megahertz, while one billion cycles per second equals one gigahertz. A 1.8 GHz processor, on the other hand, has twice the clock speed of a 900 MHz processor.

It's crucial to remember, however, that a 1.8 GHz CPU isn't always twice as fast as a 900 MHz CPU, because different chip architectures are used. For example, completing a multiplication instruction on one processor may take more clock cycles than on another processor. Other aspects play a role in a computer's overall performance. The number of CPUs, bus speed, cache capacity, RAM speed, and HDD or SSD speed are all examples. As a result, while the processor's clock speed is an important indicator of a computer's speed, it is not the only aspect to consider.

Clone

What are the similarities between sheep, droids, and computers? They're all clonable! Computer clones are by far the most prevalent of the three.

In the mid-1980s, the term "clone" was coined to designate DOS or Windows-based computers produced by firms other than IBM. "IBM clones" or "IBM compatible" computers were popular terms for the devices. The machines were dubbed clones because they worked in the same way as IBM computers. They were both powered by the same hardware and ran the exact same software.

Clones of PCs can still be found nowadays. In fact, there are now dozens more Windows-based computer manufacturers than there were in the 1980s. Many additional manufacturers, including Dell, Gateway, HP, Compaq, and Sony, produce Windows-based computers. The name "PC," which stands for "Personal Computer" in technical terms, is now commonly used to refer to IBM clones.

Apple forced the end of Macintosh clone production by having the Macintosh operating system only run-on Apple-branded machines for a few years in the late 1990s.

The term "clone" can also refer to software that performs the same function as a more well-known program. It can also refer to electrical devices that are similar to computers.

Cloud Computing

"Cloud computing allows users to back up and retrieve information over the internet," for example. Cloud computing is a broad word that refers to a variety of Internet-based services. These include social networking sites such as Facebook and Twitter, as well as online backup services and apps. Computer networks connected over the Internet for server redundancy or cluster computing are also included within the term of cloud computing.

Figure 31. Cloud computing.

Source: Image by Pixabay.

The "cloud" is a metaphor for the Internet and reflects the interconnection of internet services all over the world. Anyone with an Internet connection can utilize the cloud to store data and share it with other online users.

Cluster

A cluster can refer to two separate things in the computing world:

On a disk, a collection of sectors. A cluster is a slightly larger unit that is used to organize and identify files on your hard disk. A sector is the smallest unit that can be accessed on your hard disk. The majority of files occupy many clusters of disk space. Each cluster has its own unique ID, allowing the hard drive to locate all clusters on the disk. Some clusters may remain classified as used after reading and writing multiple files to a disk, even if they contain no data. These are known as "lost clusters," and they can be repaired with ScanDisk on Windows or Disk Utility on the Mac. This is why using a disk tool or defragmentation program can help you free up disk space.

A collection of interconnected computers. A cluster can also refer to a collection of machines that all do the same task. For example, a cluster of eight PCs connected through high-speed Ethernet could be used to handle scientific data. Because all computers in the cluster are operating as one machine, this system is commonly referred to as "parallel computing." Clusters are generally utilized for high-end processing like scientific calculations or decryption techniques.

CMOS

This technology is commonly used to make transistors and stands for "Complementary Metal Oxide Semiconductor." Unfortunately, the "complementary" component of the phrase does not imply that these semiconductors are free. Rather, it describes how they generate a positive or negative charge. CMOS-based transistors run efficiently and consume extremely little power because they only require one charge at a time. This is because the charges can remain in a single state for a long time, allowing the transistor to operate with little or no power unless it is required.

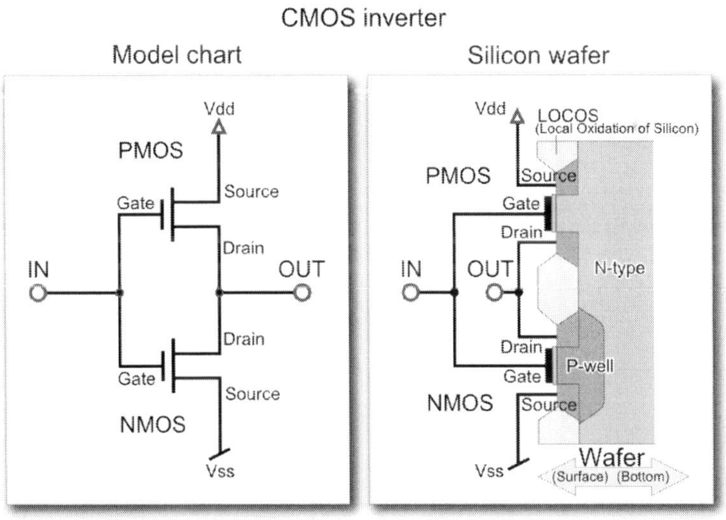

Figure 32. CMOS (complementary metal oxide semiconductor).

Source: Image by Wikimedia.

Processors that use CMOS-based transistors can run at extremely high speeds without overheating and catching fire due to their incredible efficiency. CMOS memory, which stores the date and time as well as other basic system settings, may also be found in your computer. Because of the CMOS's low power consumption, the memory can be operated for many years by a single Lithium battery.

CMYK

These are the four primary hues used in color printing. CMYK colors are "subtractive," unlike RGB (red, green, blue), which is utilized to create visuals on your computer screen. This means that as you blend the colors together, they become darker. Because RGB colors are utilized to represent light rather than pigments, the colors become brighter as you blend or intensify them.

In theory, mixing equal parts pure cyan, magenta, and yellow should result in black. True black is difficult to get by mixing the colors together due to imperfections in the inks. Therefore black (K) ink is frequently used in conjunction with the other three colors. To avoid confusion with blue in RGB, the letter "K" is used.

Codec

No, this is not a cheesy Kodak knock-off. The term "codec" stands for "coder-decoder," which is exactly what it does. Most audio and video formats employ compression to ensure that they do not consume excessive amounts of storage space. When audio and video files are saved, they are compressed with a specific codec, which is then decompressed when they are played again.

MPEG and AVI are common video codecs, whereas WAV and AIFF are common music codecs. Codecs can also be used to compress streaming media (live audio and video), allowing a live audio or video clip to be transmitted over a high-speed Internet connection.

Command Prompt

In a text-based or "command-line" interface, such as a Unix terminal or a DOS shell, a command prompt is utilized. It is a symbol or a sequence of characters at the start of a line that signifies the system is ready to accept input. To put it another way, it asks the user for a command (hence the name). The current directory of the system the user is working with is frequently displayed before the command prompt. The default prompt in

DOS, for example, may be C:, indicating that the user is working at the root level of the primary C: drive.

In Unix, the prompt could be user$, where "user" is the current user's name. The tilde () denotes the user's home folder as the current directory. At the command line, a user can execute commands like cd /, which stands for "change directory to the root folder," and allows the user to explore among multiple directories of files on a hard disk or network. A user can type dozens of other commands to list, move, delete, and copy files, start programs, and perform other tasks. While the "cd" command is the same in DOS and Unix, the syntax of many other commands differs.

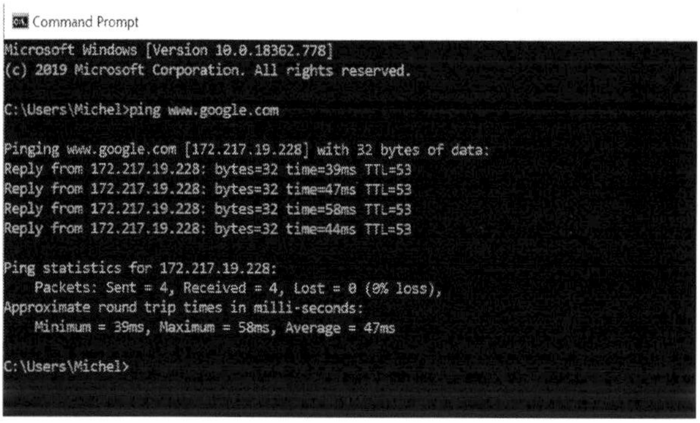

Figure 33. Command Prompt.

Source: Image by Wikimedia.

Because a command prompt demands specific input, you cannot use it if you don't know the syntax of the command you're trying to run. To use a command-line interface, you'll need to know a few fundamental commands. Fortunately, most operating systems now solicit input from the user through a graphical user interface (GUI). This means you won't need to memorize any specific commands to use the computer. Learning to utilize a command-line interface, on the other hand, is comparable to learning to drive a car with a manual transmission. You never know when it will be useful.

Commercially Available Software

There are three types of computer software: freeware, shareware, and commercial software. Freeware is free to use and does not require the user to pay anything. Shareware is also free to use, but it usually contains restrictions on the program's capabilities or the amount of time it may be used until the user pays for it. Payment is required before using commercial software, however it includes all the program's capabilities with no restrictions or time limits.

Commercial software products are usually packaged in a physical box, which you can see in stores. While software boxes aren't as large as they once were, they nevertheless include the software CD or DVD, as well as a "getting started" handbook and a registration key for activating the program. Most commercial software applications need users to register the program so that the firm can keep track of who has access to the software. Some commercial software products, such as later versions of Microsoft and Adobe programs, demand registration after 30 days in order to continue using them.

While the majority of commercial software is still sold in physical boxes, many titles are also available in the form of downloads. Typically, these downloads are made available through the company's website. Instead of receiving the software in the mail, the user pays for it directly on the website and downloads it to his computer. Paying for a registration key, which unlocks the capabilities of a shareware product, is another popular way of acquiring commercial software online. This converts the shareware program to a commercial version, which removes all of the shareware's constraints.

Compact Flash

Compact Flash is a form of flash memory that is sometimes abbreviated as "CF." Compact flash cards are typically used to store photos in digital cameras, but they are also found in PDAs and portable music players.

Figure 34. Compact Flash.

Source: Image by Flickr.

Compact Flash cards are divided into two types: Type I and Type II. Type I cards are 3.3 mm thick, while Type II are 5 mm thick. IBM makes a "MicroDrive" card that has the same dimensions as a Type II CF card, but uses an actual hard drive construction rather than flash memory. Compact Flash cards used to be able to hold only a few megabytes of data, but they can currently hold many gigabytes. The new CF+ standard has a storage capacity of up to 137 GB.

Compile

When programmers construct software, they start by writing the code in source code, which is written in C, Java, or any other programming languages. These source code files are preserved in a text-based, human-readable format that programmers may open and edit. The source code, on the other hand, cannot be run directly by the computer. The code must be translated from source code (a high-level language) to machine code before it can be recognized by the computer's CPU (a low-level language). "Compiling" the code is the term used to denote this procedure.

A compiler is included in most software development systems, and it converts source code files into machine code or object code. The resulting application is commonly referred to as an executable file because this code may be executed directly by the computer's CPU. The.EXE file extension is used by Windows executable files, while the .APP extension is used by Mac OS X apps and is generally hidden.

Compiler

A compiler is a piece of software that converts source code files into executable code. Most computer software packages contain it as part of the integrated development environment (IDE).

The compiler converts source code written in a high-level language like C, BASIC, or Java into a low-level language like machine code or assembly code. This code is tailored to a processor, such as an Intel Pentium or a PowerPC. The CPU will then detect the program and run it through the operating system.

A compiler can't change a program after it's been compiled from source code files. As a result, any modifications to the source code must be done in the source code, and the program must be recompiled. Fortunately, most current compilers can identify changes and just require recompilation of the updated files, saving programmers a lot of effort. This can help programmers cut their 100-hour work weeks down to roughly 90 hours before project deadlines.

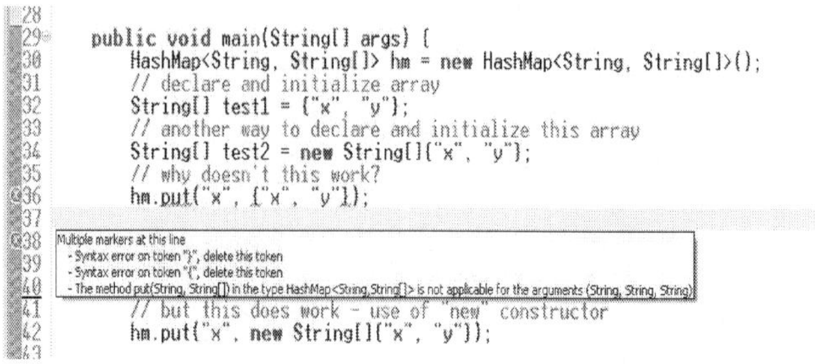

Figure 35. Compiler.

Source: Image by Flickr.

Component

A motherboard, CPU, RAM, and hard drive are only a few of the components that make up a computer. Each of these parts is built up of components, which are smaller parts.

Electrical connectors, a printed circuit board (PCB), capacitors, resistors, and transformers, for example, are all found on a motherboard. All of these pieces work together to allow the motherboard to communicate with the rest of the computer. Integrated circuits, switches, and incredibly thin transistors are all part of the CPU. These components oversee processing data and performing calculations.

A component, in general, is a subset of a larger group. As a result, computer components include the bigger portions of a computer, such as the CPU and hard drive. Components, on the other hand, are the little parts that make up these machines. Component video is a sort of high-quality video link that is sometimes referred to as component. The visual signal is sent over three independent wires in a component connection: red, green, and blue. Composite video (usually a yellow connector), which combines all of the color signals onto a single cable, providing improved color fidelity.

Configuration

When people talk about their computer setup in the computing world, they are talking to the technical specifications, or "tech specs," of their machine. Processor speed, RAM, hard drive space, and the type of video card in the machine are often included in these specifications. While there are many additional advanced specifications that you could list, most computer geeks will let you into their conversation if you know the four described above.

Figure 36. Configuration.

Source: Image by Pixabay.

Contextual Menu

When you right-click on a certain region of the screen, a contextual menu appears as a pop-up menu. Because the menu items are related to what you click on, it is "contextual." While contextual menus are not supported by all programs, they are usually a standard feature of most modern operating systems and applications.

Contextual menus offer a selection of alternatives based on where the cursor is when the right mouse button is pressed. For example, if you right-click on your computer's desktop, the menu that appears will have desktop-specific commands. Change the desktop backdrop, create a new folder, tidy up the icons, or see the desktop properties are some of the options available. You may see options to create a new folder, modify the view settings, or examine the folder properties if you right-click within a desktop window.

When you right-click within an application, you usually get menu selections that are exclusive to that program. Right-clicking on a Web browser window, for example, may bring up a menu with options like "Back," "Reload," and "Print." A list of related correctly spelled words appears when you right-click a misspelled word in a word processing tool.

In an image-editing program, right-clicking a photo may reveal a menu of modifying options. These are just a few of the many contextual menus that can be found in various programs.

A contextual menu can be thought of as a more simplified menu bar. This is because they usually have a lot of the same options as the program's main menu, but only show the items that are relevant to the spot you clicked. As a result, contextual menus can be a convenient and quick approach to choose the choice you require. Try right-clicking in different places of the screen the next time you use your favorite program. You might come upon some useful options you weren't aware of!

Control Panel

The Control Panel is a Windows operating system tool that allows users to change system settings and controls. It contains several small apps, known as control panels, that can be used to examine and modify hardware and software settings. Display, Keyboard, and Mouse settings are examples of hardware control panels. Date and Time, Power Options, Fonts, and Administrative Tools are among the software control panels.

Many control panels are built into the Windows operating system, but others can be added using third-party software or utilities. For example, if you buy a new mouse, it may come with a CD that allows you to install a control panel tailored to that mouse. Some graphics cards include an additional control panel that allows the user to fine-tune the computer's visual settings. Control panels can always be located in the Control Panel folder, regardless of when they were installed.

By going to the Start menu and selecting Control Panel, you can get to the Windows Control Panel. When you open My Computer, you'll find it in the "Other Places" portion of the window's sidebar. The Control Panel in Windows XP and Windows Vista can be seen in either Category View or Classic View. Classic View divides the control panels into sections, whereas Category View displays them all at once. While the Category View is intended to make finding different settings easier, most users who are familiar with the control panels find the Classic View to be more efficient.

Control Panels were also utilized by the Mac OS for a long time, from Mac OS 9 onwards. Control panels were unified into a single interface called System Preferences with the launch of Mac OS X. In Mac OS X, the control panels are now referred to as "Preference Panes." They can be accessed by going to the Apple menu and selecting "System Preferences" or by clicking the System Preferences icon in the Dock.

Controller Card

The controller card, or simply "controller," is a piece of hardware that connects the motherboard to the rest of the computer's components. Hard drives, optical drives, printers, keyboards, and mice, for example, all require controllers to function. All the necessary controllers are incorporated into the motherboard as chips, rather than full-sized cards, in most PCs. If you install other components, such as a SCSI hard disk, you may also need to add a controller card. Controller cards are usually inserted in one of the PCI slots of a computer.

Figure 37. Controller Card.

Source: Image by Wikimedia.

Cookie

A cookie is data given to your computer by a Web server that keeps track of your actions on a particular website. It is analogous to a preference file in a computer software. When you visit the site after receiving the cookie, it will load specific pages based on the information in the cookie.

Some websites, for example, can remember information such as your user's name and password so that you don't have to re-enter it every time you visit. Cookies are what allow you to have websites that are tailored to you. Cookies are what allow you to have personalized web sites like "My Excite" or "My Yahoo," where you can customize what is displayed on the page. While cookies provide a lot of advantages, some people don't appreciate their information is collected or monetized by the websites they visit. As a result of data privacy legislative frameworks, websites now ask you whether you accept or reject cookies.

Copy

You may copy data in many software programs, such as text in Microsoft Word or an image in Adobe Photoshop. To copy a piece of data, first select it (or highlight it) and then choose "Copy" from the program's Edit menu. The keyboard shortcuts "Control-C" for Windows and "Command-C" for Mac OS are available in most apps. When you copy data, it is saved in a buffer called the "Clipboard" in the system's memory. This is a temporary storage region in the RAM of your computer that retains the most recently copied object. Copying isn't useful if you can't use the information elsewhere. To paste the copied data into a document, go to the Edit menu and select "Paste," and the data will be pasted in.

Copyright

Copyright safeguards an author's original work from being plagiarized and republished. It grants the author exclusive publishing and distribution rights, preventing others from republishing the work. Literary works, artwork, music, video, and digital publications, such as websites and blogs, all fall under the umbrella of copyright protection. For example: "Copyright law automatically protects original websites."

Copyright protection is automatic in most nations. Copyrights may, however, be registered with a central agency for extremely valuable work. Although most websites and blogs are copyright protected by default, many include a basic copyright line near the bottom of each page. This short line of text, which frequently incorporates the "©" sign, informs visitors that the page contains copyrighted material.

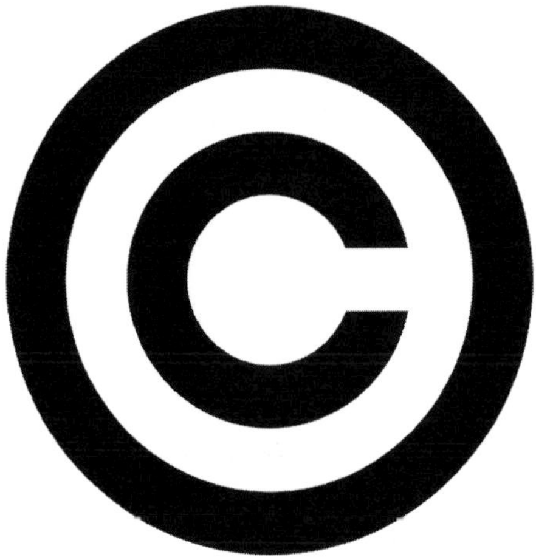

Figure 38. Copyright symbol.

Source: Image by Pixabay.

CPA

"Cost Per Action" is a term used in online advertising. When a person clicks on an advertisement on a publisher's website and then completes a specific action, the CPA is calculated. A publisher, for example, might put a banner or text link from an advertiser on his website. The user is routed to the advertiser's website when she clicks the link. She may next be prompted to complete a form or participate in a survey. The action is complete when she completes the form or survey, and the advertiser pays the publisher a specified amount based on the CPA.

The terms cost per acquisition (CPA) and cost per lead (CPL) are frequently interchanged, albeit CPA is more generic.

CPC

It stands for Cost Per Click, is a term used in online advertising. The cost per click (CPC) is the amount of money a publisher earns each time a user clicks on an advertisement link on his website. A publisher, for example, may use text or image-based adverts on his website. A visitor is led to the advertiser's website when he or she clicks on one of the advertising. The advertiser's monitoring system records each click, and the publisher is paid a certain amount based on the CPC.

CPL

"Cost Per Lead" is a term used in online advertising. CPL refers to the amount of money a publisher earns when he generates a lead for an advertising. For instance, the publisher might run an advertisement for an investing firm on his website. When a user clicks on the advertisement link, she is taken to the advertiser's website, where she can open an account with them. If she chooses to join up, a lead is created, and the publisher is compensated according to the CPL.

The terms cost per lead (CPL) and cost per action (CPA) are frequently interchanged, while CPL is more explicit.

CPM

"Cost Per 1,000 Impressions" is a term used in online advertising. The cost per 1,000 impressions of an advertisement, such as a banner ad or other promotion, is defined as a CPM. Each time an advertising is presented, an impression is recorded. While some advertisers pay publishers based on impressions alone, many marketers pay for individual clicks or leads generated by their ads. As a result, measuring pay per click (PPC) or pay per lead (PPL) rates in Web advertising may be more effective. In any case, the advertiser's goal is to produce as many leads as possible while keeping costs reasonable.

The CPM is also used by web publishers to calculate the revenue per 1,000 impressions. While technically this should be "revenue per 1,000 impressions," or RPM, from the publisher's perspective, the words CPM and RPM are frequently interchanged.

CPS

Stands for "Classroom Performance System." CPS refers to a technology system that is utilized in a classroom for educational reasons. It consists of both technology and software that work together to provide students with a modern, interactive learning environment. A projector, a CPS chalkboard, a computer running CPS software, and answer pads for students make up a typical CPS classroom setup.

The teacher can utilize a handheld CPS Chalkboard to give lessons or assessments that are graphically shown on the projector in a CPS-enabled classroom. The response pads, which are akin to remote controllers, allow students to interact with the lesson or test.

A projector, a CPS chalkboard, a computer running CPS software, and answer pads for students make up a typical CPS classroom setup. Each response pad features a number of buttons (A through H) that can be used to respond to real-time test questions displayed on the projector. Instead of everyone crying out the answer at the same time, each student can answer each question in a more logical manner.

CPS has been found to boost children's learning capacity and subject retention. CPS can dramatically improve students' interest in the subject matter because each student can interact with every individual component of the session. Given how tough it is to hold children's attention these days, CPS may be an attractive and welcome new technology for many teachers.

Cron

A cron is a procedure or task that runs on a Unix system at regular intervals, often known as a "cron job." Crons can be used to sync the time and date through the Internet every ten minutes, send an e-mail notice once a week, or back up specified directories once a month, to name a few examples. The Unix "crontab" command can be used to schedule crons. The jobs are saved in a crontab file and run by the system at predetermined intervals. of course, if you don't use Unix, you won't need to utilize a cron or even understand what one is. However, if you've read this far, you'll know that it's a great way to impress your pals.

Crop

Cropping an image or photo removes a portion of the image. This may entail chopping portions of the image from the left, right, top, or bottom, or a mix of these. It might potentially be as simple as cutting out a small rectangle from the image.

For instance, suppose a buddy takes a photo of your family and leaves a good five feet of headroom over the tallest member. You probably don't want half of the shot to be filled with sky if you're intending to use it as a Christmas card photo. As a result, you can crop the top section of the photograph with an image-editing application, leaving only a small region above the tallest person's head. You may want to crop some of the sides as well if the camera was zoomed out too far.

The majority of picture editing software includes a crop function. Normally, you would use the mouse to choose a rectangular area and then click "Crop" to eliminate the area outside the selection. After cropping the image, the clipped area is removed, leaving only the rest. As a result, if you're cropping and storing an original shot, you might wish to make a duplicate file first. Cropping is distinct from just reducing the size of an image, which entails lowering the image resolution.

Cross-Browser

"The Web developer created a cross-browser website that looked similar in Firefox and Internet Explorer," says the Web developer.

A website that is cross-browser compatible is one that can be seen in a variety of browsers. This indicates that the Web pages display properly in a variety of browsers, including Safari, Internet Explorer, and Firefox. To be interoperable across various sites, cross-browser sites may need to generate modified HTML or JavaScript.

Crossplatform

Software that can run on a variety of different computer systems is known as crossplatform software. The graphical tool Adobe Photoshop and the word processor Microsoft Word, for example, are both available for Windows and Macintosh. As a result, Photoshop and Word are considered crossplatform programs. While "crossplatform" is most commonly associated with computer software, it can also apply to hardware. Peripherals that function on both Mac and PC, such as keyboards, mouse, printers, scanners, and digital cameras, are crossplatform. Multiplatform software and hardware are software and devices that can run on many platforms.

Figure 39. Crossplatform.

Source: Image by Flickr.

CRT

CRT stands for "Cathode Ray Tube," which is the technology utilized in computer monitors and televisions. Electrons are fired from the back of the tube to phosphors in the front of the display to form the image on a CRT display. The phosphors light up when the electrons hit them, and the image is projected on the screen. The color you see on your screen is created by a combination of red, blue, and green light, known as RGB.

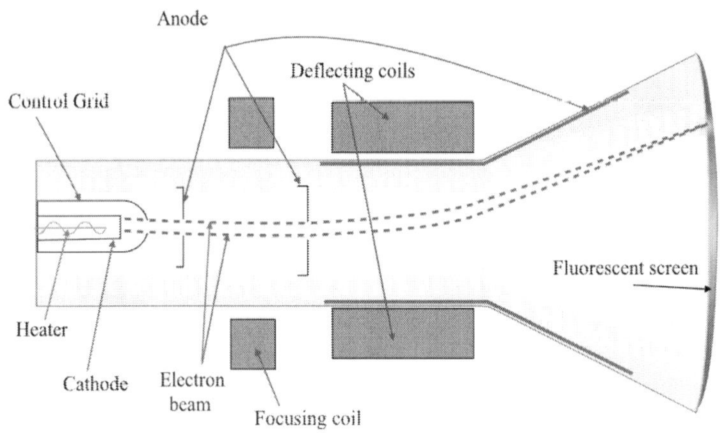

Figure 40. Cathode Ray Tube.

Source: Image by Wikimedia.

Magnetic charges guide the flow of electrons, which is why unshielded speakers or other magnetic devices put close to a CRT monitor may cause interference. Because flat screen or LCD displays do not require a magnetic charge, they do not have this issue. LCD displays also do not have a tube, allowing them to be significantly thinner than CRT monitors. As a result, flat-screen displays are rapidly displacing CRT monitors in both consumer and professional industries.

CSS

"Cascading Style Sheet" is what CSS stands for. The layout of Web pages is formatted using Cascading Style Sheets. They can be used to define text styles, table sizes, and other characteristics of Web pages that were previously solely defined in the HTML of the page CSS enables Web designers to establish a consistent look across multiple pages of a website.

Commonly used styles only need to be declared once in a CSS document, rather than describing the style of each table and each block of text within the HTML of a page. Once a style is defined in a cascading style sheet, any page that mentions the CSS file can utilize it. CSS also makes it simple to update the style of multiple pages at once. For 50 pages of a website, a Web developer could want to increase the default text size from 10 pt to 12 pt. If all the pages use the same style sheet, changing the text size on the style sheet is all that is required, and the larger text will appear on all of the pages.

While CSS is best known for producing text styles, it may also be used to format other components of a Web page's layout. CSS can be used to determine table cell padding, the style, thickness, and color of a table's border, and the padding around images and other objects, for example. CSS allows Web designers more precise control over the appearance of Web pages than HTML does. Therefore, most modern Web pages use cascading style sheets.

CTP

"Composite Theoretical Performance" is what CTP stands for. Computer processor performance is measured using CTP. CTP calculations return values that are used in benchmarking to compare the performance of different processors. Intel and AMD, for example, use CTP calculations to determine how many millions of notional operations per second are performed (MTOPS) An Intel Pentium M 770 processor with a clock speed of 2.13 GHz has a CTP of 7100 MTOPs, while an AMD Opteron 146 processor with a clock speed of 2.0 GHz has a CTP of 7168 MTOPS.

Faster processor speeds may not usually imply a greater CTP, as shown in the example above. Other factors, such as the architecture of the processor and the speed of the frontside bus, have an impact on total performance. CTP can be used to compare different brands of processors as well as different models of the same company's processors. CTP is now being used to quantify the performance boost when numerous processors are used simultaneously, as dual, and quad processors become increasingly common.

CTP also stands for "Computer to Plate." This is a method of creating printing plates without the need of expensive film. Rather, the images are transferred directly from the computer to the plate.

CTR

CTR stands for "Click-Through Rate," and it's a term that's commonly used in online advertising. When a visitor clicks on one of the advertising on a website, the publisher normally receives money from the advertiser (the PPC model). As a result, publishers are curious as to what is the percentage of page visits which result in ad clicks. The higher a publisher's click-through rate is, the more money he or she will make. So, what do you think a typical website CTR is? 5% vs. 10% vs. 5% vs. 10% vs. 1% is a good start. Because the average click-through rate (CTR) on the internet is roughly 1%, publishers must attract thousands of visitors each day to gain significant money.

Cursor

The cursor on your screen can represent two things: (1) the location of your mouse pointer, and (2) the location of the next character typed in a line of text.

Most of the time, the mouse cursor is an arrow that you can use to point to various objects on your screen. You can conduct an action on an object by clicking or double-clicking the mouse button when the cursor is over it (such as opening a program). When you roll over a link in a Web page, the mouse pointer can transform into a little hand, or an hourglass (when Windows is "thinking" so hard that it won't let you click on anything).

A text cursor is usually a vertical line or an I-shaped item that flashes in the middle of a line of text. Because you're adding new text to the uncharted white region of the page, the cursor will usually be near the end of the line when you're composing a paper.

You can use the mouse pointer to click the spot where you wish to insert the text if you want to put a word or phrase somewhere else in a line of text. When you start typing in most word processing systems, the text cursor continues to flash, but the mouse pointer vanishes until you move the mouse again. Because most people can't type and click at the same time, this is done to avoid "cursor confusion."

Cut

The Cut command copies a selected object to the clipboard after removing it from the screen. It conducts a simultaneous Delete and Copy operation. Data can be placed into another similar document after it has been cut from one. Text, photos, audio, and video selections can all be chopped. However, because the command removes the selection from the page, only editable elements can be clipped.

Cyberbullying

"You should report cyberbullying to a parent or teacher if you are a victim of cyberbullying." Cyberbullying is when a young person is harassed, humiliated, or threatened online. It incorporates within it a variety of communication channels, including the Internet, e-mail, and cell phone conversation. Cyberbullying can occur in a variety of ways.

Posting nasty comments or photos of another person online, sending sexual text messages, or bombarding a person with unsolicited e-mails are just a few instances. It could also entail forming a group to target a single individual and making defamatory comments on social networking sites such as Facebook and MySpace. When it comes to adults, cyberbullying is usually referred to as cyber-harassment or cyberstalking.

Figure 41. Cyberbullying.

Source: Image by Wikimedia.

Cybercrime

Cybercrime is defined as criminal action carried out using computers and the Internet. This might range from stealing millions of dollars from internet bank accounts to downloading illicit music downloads. As a result, it's a good idea to double-check a site's URL or Web address before providing personal information.

Because cybercrime encompasses such a broad range of illegal conduct, the examples above are just a handful of the hundreds of offenses that fall under this umbrella. While computers and the Internet have made our lives easier in many ways, it is terrible that some people take advantage of others by using these tools. As a result, it's a good idea to safeguard yourself by installing antivirus and spyware-blocking software and being cautious about where you submit personal information.

Figure 42. Cybercrime has crossed international borders.

Source: Image by Pixabay.

Cyberspace

Unlike most computing words, "cyberspace" lacks a universally accepted definition. Instead, it's used to describe the computer's virtual world. A block of data floating through a computer system or network, for example, is referred to as an object in cyberspace. With the introduction of the Internet, cyberspace has expanded to include the entire global computer network. So, after sending your friend an e-mail, you may declare you communicated with her over cyberspace. However, this term should be used with caution, as it is a popular novice term that has been abused.

The term "cyberspace" is credited to William Gibson, who coined it in his 1984 novel Neuromancer. As defined by Gibson, cyberspace is "a collective delusion shared by billions of authorized operators in every country, as well as students learning mathematical principles… Data extracted from the banks of every computer in the human system is represented graphically. Complexity beyond comprehension. Lines of light, clusters, and constellations of data spanned the mind's non-space "Berkley Publishing Group, New York, 1989, 128.

Cyber squatters

"Cybersquatters frequently register hundreds or even thousands of domain names," says one cybersquatter. A cybersquatter, often known as a "domainer," is someone who registers a domain name but does not plan to use it for a website. Many cybersquatters also set up parked pages for their domain names, which can bring in money through advertising. While cybersquatters often profit from the sale of domain names, they may be forced to relinquish control of others. Cybersquatters are prohibited from denying domains to individuals or businesses who have previously trademarked or registered similar names under the Anti-Cybersquatting Consumer Protection Act (ACPA).

D

DAC

The abbreviation DAC stands for "Digital-to-Analog Converter," and it's commonly called "dac." Computers can only recognize digital data, hence the output they create is usually in digital format. However, some output devices only accept analog input, which necessitates the use of a digital-to-analog converter or a digital-to-analog converter.

The most common use for a DAC is to convert digital audio to an analog signal. This conversion typically takes place in the sound card, which has a built-in DAC. The digital signal, which is basically a stream of ones and zeros, is transformed into an analog signal that might take the form of an electrical charge. This electrical charge is recognized by most speaker inputs and therefore can be output to a speaker system.

DACs are also used for converting video signals. Historically, most video displays, such as TVs and computer monitors used analog inputs. Only in the last couple of years have digital displays with DVI and HDMI connections become commonplace. A DAC is most commonly used to convert digital audio to analog audio. The sound card, which has a built-in DAC, is usually where this conversion happens. Most speaker inputs recognize this electrical charge, so it can be sent to a speaker system.

Video signals are also converted using DACs. Most video displays, such as televisions and computer monitors, used analog inputs in the past. Digital displays using DVI and HDMI connectors have just become mainstream in the last several years. As a result, a computer's digital video signal must be converted to an analog signal before it can be output to an analog display. This is why a DAC is included with all video cards that have an analog output (such as a VGA connection).

There is a risk of quality loss when converting a signal from one format to another. As a result, whether you're converting audio or video signals, a high-quality DAC is essential. Performing the opposite conversion, which requires an analog-to-digital converter, or ADC, is the same.

Daemon

The word "daemon" is derived from the Greek language and refers to a "inner or accompanying spirit" (Oxford American Dictionary). A computer daemon is a constantly running program that activates activities when it gets particular input, therefore this is a fitting name.

When a user desires to print a document, for example, a printer daemon spools information to a printer. On a mail server, a daemon routes incoming mail to the relevant mailboxes. When users browse Web pages, web servers employ the "HTTPD" daemon to transfer data to them. Daemons were first employed by the Unix operating system, but they have now been included into the Unix-based Mac OS X.

Dashboard

With the release of Mac OS X 10.4 Tiger, Apple introduced Dashboard, a user-interface component. It provides access to a variety of "widgets" that display information such as the time, weather, stock prices, phone numbers, and other pertinent information. Apple includes widgets that do all of these tasks, as well as a calculator, language translator, dictionary, address book, calendar, unit converter, and iTunes controller, with the Tiger operating system. Aside from the standard widgets, there are hundreds of third-party widgets available that allow users to do things like play games, check traffic, and view sports scores, to mention a few.

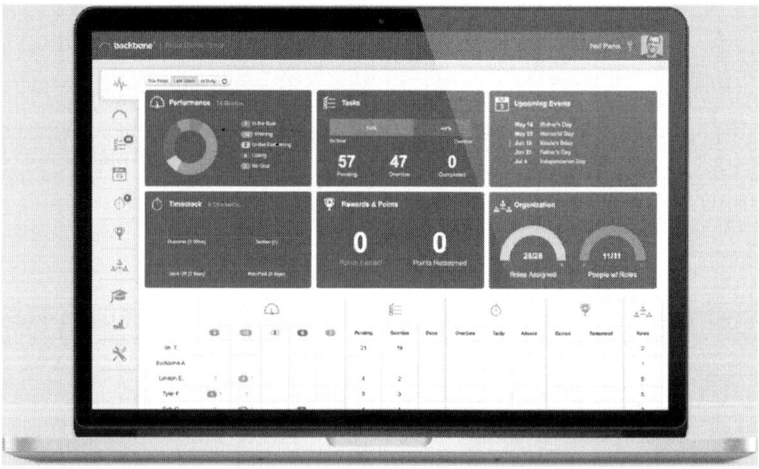

Figure 43. Dashboard.

Source: Image by Backbonepro.com.

By selecting the Dashboard application, you can view the widget dashboard. The user can see a list of all installed widgets by clicking the plus "+" symbol in the lower-left corner of the screen. The widgets become active when you click or drag them onto your desktop. They can be closed individually by clicking the close box. By using the F12 keyboard shortcut, they vanish until the user needs them again.

Data

Computer data refers to information that has been processed or saved by a computer. Text documents, photos, audio recordings, software applications, and other sorts of data may be used to store this information. Computer data is saved in files and directories on the hard disk and may be processed by the computer's CPU.

Computer data is a collection of ones and zeros, known as binary data, at its most basic level. Computer data may be created, processed, saved, and stored digitally since it is in binary format. Data can be moved from one computer to another via a network connection or a variety of media devices. It also doesn't degrade or lose quality over time when used several times.

Data Management

"The company's data management has been delegated to the systems administrator." The way individuals, businesses, and organizations manage computer data is referred to as data management. It encompasses both micro and macro applications, such as data architecture and design, as well as data storage, access, and security.

Although computer data is intangible, it is still precious. As a result, all users should think about how they maintain their data. This could include things like backing up vital files and encrypting personal data. Rate of Data Transfer, the data transfer rate is a typical metric for determining how quickly data is moved from one point to another. A hard disk, for example, may have a maximum data transfer rate of 480 Mbps, yet your ISP may only offer a maximum data transfer rate of 1.5 Mbps.

Figure 44. Data management is the fundamental task in computer science.

Source: Image by Flickr.

Data transfer

Data transfer rates are usually expressed in bits per second (bps) rather than bytes per second (bps), which might be confusing. Because a byte has eight bits, an 80 Mbps sustained data transfer rate only transfers 10MB per second. While this may be perplexing for consumers, Internet service providers must appreciate monitoring data transfer rates in bits per second (bps) because it makes their Internet access speeds appear 8 times quicker than they actually are.

Data type

Data type is a data storage format that can hold a particular kind or range of values.

When data is stored in variables in computer programs, each variable must be allocated a specific data type. Integers, floating point numbers, characters, strings, and arrays are some of the most popular data types. Dates, timestamps, Boolean values, and varchar (variable character) formats are examples of more particular kinds.

Before assigning a value to a variable, certain computer languages require the programmer to declare its data type. When initial data is entered into a variable, other languages can up-automatically assign the variable's data type. If the value "1.25" is entered into the variable "var1," the variable is formed as a floating-point data type. The variable would be assigned a string data type if it was set to "Hello world!" Most programming languages allow a single data type to be stored in each variable. As a result, if the data type of the variable has already been changed to integer, assigning string data to the variable may result in the data being converted to integer format.

Database

Database applications also make use of data types. It consists of one or more tables, each with multiple fields. A data type, such as string or integer, is assigned to each field. Each record that is entered into the database has data in these fields. The information in a database can be quickly searched, sorted, and updated since it is stored in ordered records.

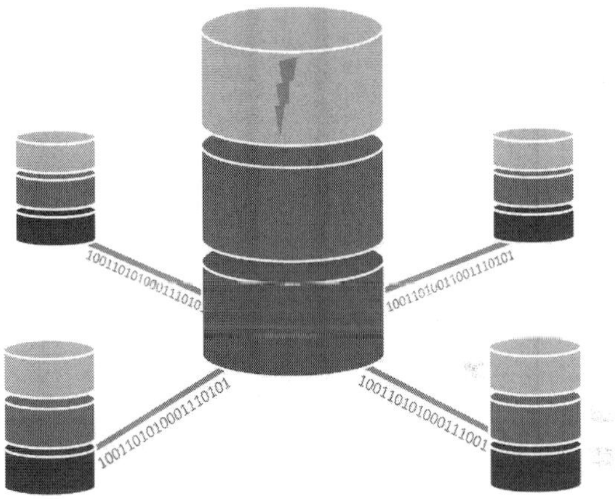

Figure 45. Database.

Source: Image by Pixabay.

DBMS

Stands for "Database Management System." In a nutshell, a database management system (DBMS) is a database program. It is a software system that uses a standard approach of classifying, retrieving, and conducting queries on data in a technical sense. The DBMS organizes and manages incoming data, as well as allowing users and other programs to modify or extract it.

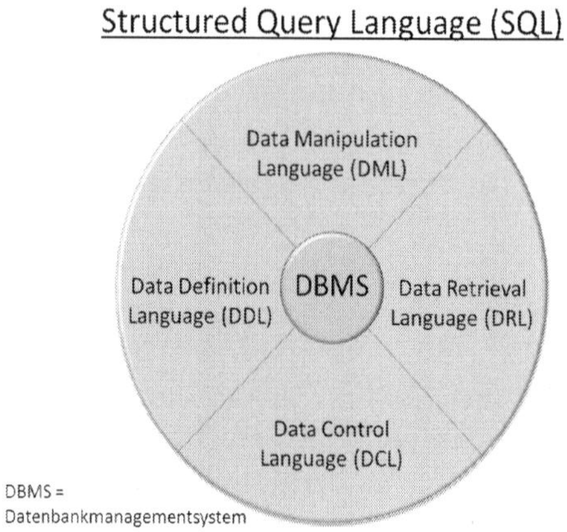

Figure 46. Database management system.

Source: Image by Wikimedia.

MySQL, PostgreSQL, Microsoft Access, SQL Server, FileMaker, Oracle, RDBMS, dBASE, Clipper, and FoxPro are some examples of database management systems. Because there are so many database management systems out there, it's critical that they can connect with one another. As a result, most database software includes an Open Database Connectivity (ODBC) driver, which enables the database to communicate with other databases. Common SQL instructions like SELECT and INSERT, for example, are converted from a program's proprietary syntax into a syntax that other databases understand.

DDL

The acronym DDL stands for "Data Definition Language." It's used to describe a database's structure, including the tables, columns, and data types it includes. The Structured Query Language, a subset of SQL, is the most widely used DDL. DDLs, on the other hand, can be used to describe a variety of data structures, such as XML schemas and other data models. Example "The DDL includes a "create" command that can be used to create a new database or table."

DDR

"Double Data Rate" is the abbreviation for "Double Data Rate." It's a more advanced kind of SDRAM, which is a sort of computer memory. DDR-

SDRAM, often known as "SDRAM II," can transport data twice as quickly as traditional SDRAM chips. This is due to the fact that DDR memory is capable of sending and receiving signals twice every clock cycle. DDR-efficient SDRAM's functioning makes it ideal for notebook PCs because it consumes less power.

DBR 2

It stands for "Double Data Rate 2." DDR2 RAM is a speedier and more energy-efficient variant of DDR memory. DDR2 memory may deliver data on both the rising and falling edges of the processor's clock cycles, just like normal DDR memory. The amount of work the RAM can do in a given period of time is practically doubled. DDR and DDR2 are both form of SDRAM, allowing them to run at a higher rate than standard memory.

While DDR and DDR2 RAM are comparable in many ways, DDR2 RAM is designed differently than DDR memory. DDR2 RAM is faster than conventional DDR memory because of its improved architecture. The RAM's bandwidth has been increased because of the new design, allowing more data to run through the RAM chip at once. This improves the memory's efficiency. Even though DDR2 memory is quicker, it uses less power since it runs more effectively than normal DDR memory. DDR2 memory's only drawback is that it is incompatible with normal DDR slots. Before updating your memory, ensure sure your machine supports DDR2 RAM.

DDR3

DDR3 (Double Data Rate Type 3) is an SDRAM specification that was first introduced in 2007. DDR3 utilizes 30% less power and gives much better data transfer speeds than the preceding DDR2 standard.

DDR3 memory modules come in both DIMM and SO-DIMM configurations. Desktop PCs normally use DIMMs, although laptops and all-in-one computers typically use SO-DIMMs. Even though DDR3 DIMMs and SO-DIMMs are the same size as DDR2 DIMMs and SO-DIMMs, they are incompatible with DDR2 RAM slots. Fortunately, the connecting connectors are positioned differently, thus inserting a DDR3 memory module into a DDR2 RAM slot is physically impossible, and vice versa.

Several types of DDR3 memory exist, providing a range of maximum data transfer rates. The first version of DDR3 memory, called DDR3–800 (also "PC3–6400") provides peak transfer rates of 6,400 MB/s. Other specifications include DDR3–1066, DDR3–1333, DDR3–1600, DDR3–

1866, and DDR3–2133. DDR3–2133 (also called PC3–17000) supports peak transfer rates of 17,067 MB/s.

Debug

Like everyone else, computer programmers aren't flawless. This means that the programs they produce may have minor flaws known as "bugs." These flaws can range from the minor errors, such as a failure to recognize user input, to the more significant errors, such as a memory leak that causes the program to crash. Programmers "debug" their programs before exposing them to the public, removing as many mistakes as possible. This debugging procedure can take a long time because resolving one issue can lead to the emergence of new ones. It's far easier to debug your windshield at a petrol station than it is to debug a computer program.

Figure 47. Debug.

Source: Image by Thenounproject.com.

Debugger

Even the most seasoned software developers rarely get it right on the first try. Certain flaws, commonly referred to as bugs, can occur in programs, leading them to fail to work as intended by the programmer. Some problems are simple to rectify, while others are quite difficult to track out. This is particularly true for huge programs with thousands of lines of code.

Fortunately, debuggers are programs that assist software engineers in finding and eliminating flaws while building programs. A debugger informs the programmer about the types of mistakes it discovers and, in many cases, highlights the lines of code where the errors occur. Debuggers also allow programmers to execute a program step by step in order to figure out when and why it breaks. Advanced debuggers give precise information about the program's threads and memory usage at each stage of execution. A powerful debugger program is comparable to OFF! With 100% deed.

Default

A pre-set value for some option in a computer program is referred to as this word. When the user hasn't given a value for a setting, this is the default value. The default typeface in Netscape Communicator, for example, is "Times." The "Times" typeface will be used by default until you go to Netscape settings and change it to something different. Default settings are usually set to what most people would prefer, so there's rarely a cause to change them. If you're one of those people who wants to tweak everything, you can modify all of the default settings you desire.

"Default" can be used as a verb as well. If a custom setting fails for some reason, the application "reverts" to the default configuration. Let's imagine you're working on a networked computer and you print anything even though no printer is mentioned. If you're lucky, the print job will default to the default printer and your work will be printed.

Defragment

Defragmenting your hard drive is an excellent technique to improve your computer's speed. Though the word "defragment" has a negative connotation, it is a straightforward and beneficial procedure. After all, a happy hard disk is one that has been defragmented.

It is a frequent operation to add and delete data from your hard drive. Regrettably, this procedure is not always carried out efficiently. When you delete a collection of little files and replace them with a single huge one, the file may be split into multiple pieces on the hard disk. The newly added file will still be read by the computer as a single legitimate file, but the drive will have to scan numerous sections of the disk to read it. Because one of the most major bottlenecks in a computer's performance is hard disk seek time, this can significantly slow down your computer. You might hear extra grinding, sputtering, and other strange noises coming from your computer if you have a lot of "fragmented" files on your hard disk.

Fragmented files irritate your computer just as much as they irritate you. This is why defragging your hard drive is so important. It's time to defragment if you're hearing extra grinding sounds or your computer isn't opening files as rapidly as it used to. You may defragment your hard disk using Windows by using the pre-installed Intel defragment tool.

Degauss

Have you ever wondered what the "degauss" button on your display does other than make a buzzing noise and make the screen go crazy for a second? Despite its primary use, the degauss button has another beneficial benefit. You must first realize that the earth contains natural magnetic fields in order to comprehend it. Magnetic charges from these fields can accumulate inside your monitor, causing color accuracy to deteriorate. Degaussing removes the monitor's negative magnetism and replaces it with positive karma. If your monitor doesn't have a degauss button, don't worry; many new monitors do so automatically.

Del.icio.us

Del.icio.us, pronounced simply "delicious," is a social bookmarking website where users may save and share Web pages they come across. Other people frequently bookmark Web pages that they find in other users' bookmarks since their bookmarks are made public and available by other users. Del.icio.us tracks how many people bookmark each site and displays the most popular ones on its home page.

News items, online learning resources, and tech assistance pages are among the most popular Del.icio.us bookmarks. The greatest Web pages eventually ascend to the top of the popularity chain as other users add beneficial pages to their bookmarks. The end product is a collection of Web pages that are helpful and worthwhile visiting. Del.icio.us allows users to browse or search its bookmark database.

When a user saves a bookmark, he or she can include a description as well as tags (keywords) relevant to the Web page. This improves the page's visibility in relevant searches. Because the sites have all been picked by users, the results of Del.icio.us searches are often of greater quality than those of a standard search engine. Other Web users appear to agree, as the website became so popular that Yahoo! Purchased Del.icio.us at the end of 2005.

The domain name "del.icio.us" is a creative modification of the standard domain syntax, where "del" is the domain prefix and "icio.us" is the domain name, with "us" being the domain suffix.

Delete

Delete is computer jargon for "delete" or "erase." You can erase entire files or directories from your hard drive, as well as text from a document. You can delete characters below the cursor when typing a document by using the delete key. You can delete characters in front of the cursor by pressing the tiny delete key on the keyboard near the home and end buttons. You can even delete entire parts of text by selecting the text and pressing either the delete or backspace keys on your keyboard.

Drag files and folders to the Recycle Bin (Windows) or the Trash (Macintosh) and empty the trash to eliminate them from your hard drive. When you delete a file, the reference to the file is destroyed rather than the file itself. This means that deleted files are still accessible until they are overwritten. Special programs, such as Norton Unerase, can retrieve files that have been mistakenly destroyed.

Desktop

The desktop on your computer is very similar to a physical desktop. Pens, papers, folders, and other often used objects are likely to be kept on your desk. The desktop on your computer serves the same purpose: to provide quick access to files on your hard drive. On your desktop, it's customary to keep commonly used files, directories, and apps. Instead of rummaging through the director, you can reach the desired objects quickly this way.

The desktop is a fundamental feature of both the Macintosh and Windows user interfaces. Both operating systems allow you to drag and drop items onto and off the desktop, as well as organize and clean up the items on the desktop. Yes, it would be great if there was a similar option for a real-world desktop. You can also personalize your computer's desktop by selecting a pattern or background image.

Desktop Computer

A desktop computer is a computer system that is designed to be used at a desk or table. Some desktop computers feature a separate monitor and system unit, while others are "all-in-one" devices with the monitor built-in. System units are normally positioned on the ground, while all-in-one PCs are built to sit on a desk. Input devices on both types of desktop computers are a keyboard and mouse.

Desktop computers, unlike laptops, are not portable devices. Instead, they're made to stay in one place and need to be connected to a power source. Desktop computers are the most common home and commercial workstations due to their static nature.

Desktop Publishing

Documents and photos are "published" when they are printed. Prior to the introduction of computers, the publishing process necessitated the use of enormous print presses that copied and replicated pages. Text and graphics would have to be printed separately, cut out, arranged on a single sheet, taped in position, then copied and printed to print images and words on the same page. Thankfully, computers with graphical user interfaces have made desktop publishing possible, allowing this procedure to be completed electronically.

Desktop publishing is defined as the use of a computer to create a printable document. The term is most generally used, however, to refer to professionals.

Desktop publishers develop page layouts for content they want to print using applications like Adobe InDesign and QuarkXPress. Books, magazines, newspapers, flyers, pamphlets, and a variety of other printed publications can all be created with these desktop publishing programs. Publishers can also produce printable visuals with applications like Adobe Photoshop and Illustrator. Basic desktop publishing can be done with word processing apps like Microsoft Word.

Complete desktop publishing entails typesetting (text layout and font selection), graphic design, page layout (how everything fits on the page), and document printing. Desktop publishing, on the other hand, might be as basic as typing and printing a school paper, as previously said. You only need a computer, monitor, printer, and software that can create a printable document to desktop publish. While this is more expensive than a pen and paper, it is significantly less expensive and more economical than a printing press!

DHCP

It stands for "Dynamic Host Configuration Protocol." This protocol is used by a network server to dynamically assign IP addresses to networked machines. When a machine connects to the DHCP server, it is assigned an IP address from a master list maintained on the server. Because IP addresses do not have to be manually issued to each machine on the network, DHCP aids with the setup of big networks. DHCP is the most widely utilized networking protocol because to the smooth automation involved.

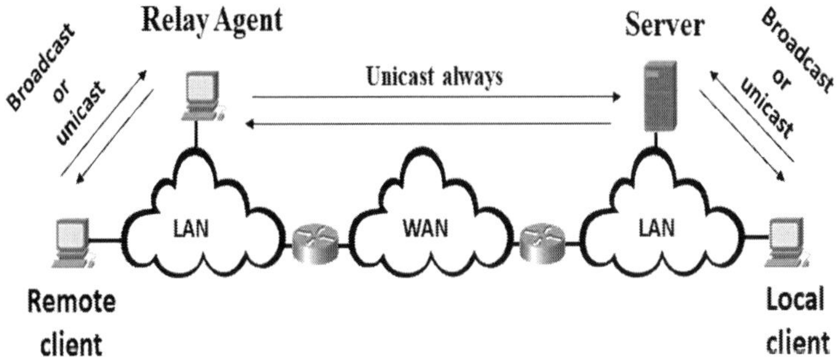

Figure 48. Dynamic Host Configuration Protocol.

Source: Image by Wikimedia Commons.

Dial-up

"Before upgrading to DSL, the family used a dial-up connection to access the Internet." A dial-up connection connects to an ISP or another computer using a modem. It transfers data at up to 56 kbps over conventional analog phone lines. Dial-up was the only means to connect to the Internet before the year 2000. Many users currently access the Internet through a DSL or cable modem connection. Both cable and DSL provide a consistent connection and data transmission speeds that are more than 100 times quicker than dial-up modems.

Dialog box

A dialog box, as its name suggests, is used to start a conversation with the user. It's a pop-up window on the screen providing options for the user to choose from. After choosing their option, the user can either click "OK" to save the modifications or "Cancel" to discard them. When a menu choice with an ellipsis at the end, such as "Preferences…" or "Save As…," is selected, a dialog box is usually opened.

If a user chooses "Internet Options…" from the Internet Explorer Options menu, for example, a dialog box will appear, allowing the user to alter the default home page, adjust the security settings, clear the browser cache, and change a variety of other options. After making their choices, the user can either click "OK" to use the new settings or "Cancel" to undo them. The "Apply" option in some Windows programs activates the options without closing the dialog box.

An "Open dialog box" shows when a user selects "Open..." from the File menu, allowing the user to browse the hard drive and other drives for files to open. When the user selects "Save As..." from the File menu, a "Close dialog box" appears, allowing the user to type the file's name and save location. While dialog boxes may not appear interesting, they are an important feature of today's computer interfaces because they give an intuitive way to connect with the computer.

Digital

A string of ones and zeros is used to store digital data. Because computers can only read information as on or off—1 or 0, they are digital machines. Although this method of processing, also known as the binary system, appears to be simple, it may be used to represent enormous amounts of data. Even though CDs and DVDs are made solely of ones and zeros, they may be utilized to store and play high-quality audio and video.

Humans, unlike computers, perceive information in analog form. They record a continuous stream of audio and visual inputs.

Digital instruments, on the other hand, use ones and zeros to estimate this information. The accuracy of the digital estimation is determined by the rate of this estimation, known as the "sampling rate," as well as the amount of information included in each sample (the bit depth).

A standard CD audio track, for example, is sampled at 44.1 KHz (44,100 samples per second) with a 16-bit bit depth. This yields a high-quality analog audio signal estimation that sounds natural to the human ear. A higher-quality audio format, such as a DVD-Audio disc, might sample at 96 KHz with a bit depth of 24 bits. On the DVD-A, the same song played on both discs will sound smoother and more dynamic.

An analog signal is more accurate than a digital signal since digital information simply guesses analog data. Computers, on the other hand, can only deal with digital data, thus storing data digitally makes more sense. Digital data, unlike analog data, may be duplicated, modified, and relocated without losing quality. Digital information has become the most frequent technique of storing and reading data due to the advantages it provides.

E

E-commerce

Business conducted over the Internet is referred to as e-commerce (electronic-commerce). Amazon.com, Buy.com, and eBay are all examples of e-commerce websites. Business-to-Consumer (B2C) and Business-to-Business (B2B) are the two main types of e-commerce (B2B). While Amazon.com caters primarily to consumers, other companies specialize in providing goods and services to other businesses. The terms "e-business" and "e-tailing" are often used synonymously with e-commerce. They both refer to the same concept; the difference is that they are intended to confuse those learning computer jargon.

Figure 49. E-commerce: Business on Internet.

Source: Image by Pixabay.

E-mail Bankruptcy

Most of us receive multiple e-mails per day in this age. Depending on your employment, you may receive dozens of non-spam mails per day. While keeping up with the multitude of e-mails received in a single day is difficult enough, falling behind a few days might be practically impossible to catch up on. After a while, you may find yourself with hundreds of unanswered messages in your inbox.

If you've become engulfed in an endless stream of e-mail in your inbox, declaring e-mail bankruptcy may be your only option. E-mail bankruptcy, like financial bankruptcy, is wiping off losses and beginning again.

The most diplomatic method to declare e-mail bankruptcy is to combine all the e-mail addresses from the unanswered messages into a single message. Then send a message apologizing for not responding and explaining that you have fallen behind on your e-mail. The quicker, but less considerate, option is to delete all of the previous communications and start over as if nothing had happened.

While keeping up with your e-mail is the greatest method to avoid e-mail bankruptcy, for other people it may be the only option to stay current with their correspondence. If you're feeling overwhelmed by the growing number of messages in your inbox, make sure you respond to the most important messages first.

Email

It's difficult to recall what life was like before e-mail. E-mail has become one of today's basic modes of communication, ranking alongside the Web as one of the most useful elements of the Internet. Every year, billions of messages are transmitted.

E-mail is a protocol that is part of the TCP/IP standard. SMTP (Simple Mail Transfer Protocol) is used to send messages, and POP3 (Post Office Protocol 3) or IMAP (Internet Message Access Protocol) is used to receive them (Internet Message Access Protocol) IMAP is a modern technology that lets you read and sort messages on the mail server without having to download them to your computer. Even though e-mail was originally designed to convey simple text messages, it has become more robust in recent years.

HTML-based e-mail may now combine styled text, colors, and images into the message using the same coding as Web sites. Documents can also be attached to e-mail messages, allowing files to be sent over the internet. However, because e-mail was not built to handle huge file transfers, most mail servers will not enable you to send large documents (over 3 MB, for example). So, keep your attachments small.

Emoji

Emojis are little images that can be put into text in the same way that ordinary characters can. Emojis first gained popularity in 2010 when it was used to depict emotions (☺, ☹, etc.). More emojis were created as the popularity rose, and they now represent a wide range of things, including people, locations, and items. Emojis also help interpret message tone and context. A recipient of the message may say: "Because you didn't use an emoji, I wasn't sure if you were serious or not."

Figure 50. Emoji.

Source: Image by Thenounproject.com.

Emojis are currently supported by most modern operating systems, including Windows, macOS, iOS, and Android. An "emoji keyboard," which is an on-screen keyboard that appears on a computer or mobile device, can be used to insert an emoji.

Your smartphone generates a Unicode code when you write an emoji, just like it would a letter, integer, or symbol. There is a binary code for an "airplane" emoji, or, just as there is a code for the letter "L."

Because the Unicode character set standardizes the list of available emojis, the emojis you can use are determined by your device and operating system version. If your device doesn't support a certain emoji, it will appear as an empty box (). When you upgrade your device's software, you may be able to see (and use) new emojis that weren't previously available.

Important: The term "emoji" comes from the Japanese words "e" (絵) and "moji" (文字), or "picture character."

Emoticon

These are the text-based faces and objects you encounter in e-mail and internet chat all the time. They assist the reader in getting a sense of the writer's emotions behind the writing. The traditional =) face, for example, indicates that the writer is pleased with something or that his message is humorous. The =P expression is used to express dissatisfaction or to indicate "Whatever…" Emoticons can be used to make real-life items as well. A @-----, for instance, is a long-stemmed rose that can be used to express affection.

Emulation

Emulation is derived from the verb "emulate," which means "to imitate or reproduce." As a result, computer emulation is the process through which one system imitates or reproduces another. This can be accomplished using hardware, software, or a mix of both. However, because hardware is costly to replicate, most of the emulation is done through the use of software.

Running various operating systems in a virtual environment is one of the most prevalent types of software emulation. Windows and other operating systems can be run on an Intel-based Macintosh computer using tools like Parallels Desktop, VMware, and Apple's Boot Camp. Sun Microsystems' xVM VirtualBox allows multiple operating systems to be run on Windows, Mac, and Unix platforms. These applications are collectively referred to as software emulators, since they emulate different computer systems.

Another common sort of software emulation allows console video games, such as those from Nintendo, Sega, and Sony, to be played on a computer. ZSNES, for example, lets you to play Super Nintendo (SNES) games on a Windows or Unix PC. Users can play Game Boy Advance games on Windows or Macintosh computers using the Visual Boy Advance emulator. ROM files, which are precise duplicates of a game cartridge or disk, are used to save games for these emulators. As a result, video game emulators load ROM files from the computer's hard drive rather than a physical cartridge.

A disk image is a sort of file used in a third form of emulation. Disk images can be used to simulate optical media and actual hard disks, whereas ROM files are used to emulate game cartridges. An .ISO (Windows) or .DMG (Mac) file, for example, can be mounted on the desktop, causing the operating system to recognize the file as a real drive. This makes making precise copies of CDs, DVDs, and hard disks simple. Disk images are also frequently used to construct user-unmodifiable virtual installer disks.

Encryption

Encryption is the process of coding or scrambling data such that only those with the correct decoding key can decode and read it. Encryption is used to keep data safe. Encryption is utilized in secure Web sites and other data transport methods. They wouldn't be able to access the information you supplied through an encrypted connection if it was intercepted by a third party. So, if you're sending a message to a co-worker over the workplace network about how much you despise your job, your boss, and the entire company, it's a good idea to transmit it over an encrypted connection.

Figure 51. Encryption: A process of encoding information.

Source: Image by Pixabay.

End User

A software application or hardware gadget is developed for a specific person. The word comes from the idea that a software or hardware product's "ultimate objective" is to be beneficial to the user. The end user might be compared to the product's developers or programmers. The installers and administrators of the product are likewise separated from the end consumers.

To put it another way, an end user is someone who utilizes software or hardware after it has been fully produced, marketed, and installed. It's also the one who continually calling the "IT guy" asking why the product isn't working properly. The phrases "user" and "end user" are interchangeable in most cases.

Ethernet

To be clear, the first syllable is pronounced "eath" in the sense of "Heath Bar," not "eth" in the sense of Bethany. In a local area network, Ethernet is the most popular type of connection (LAN). The appearance of an Ethernet port is similar to that of a standard phone jack, except it is slightly wider. This port can be used to link your computer to another computer, a local network, or a DSL or cable modem that is external

Ethernet is divided into two types: 10BaseT and 100BaseT. Through a copper line, data transfer speeds of up to 10 MBps (megabits per second) can be achieved in a 10BaseT Ethernet connection. Transfer speeds over a 100BaseT Ethernet connection can reach 100 MBps. There is also a new technology known as "Gigabit" Ethernet, which allows for data transfer rates of up to 1000 megabits per second. That's lightning fast.

Figure 52. Ethernet is used in local, metropolitan, and wide area networks.

Source: Image by Flickr.

Excel

Microsoft Excel is a spreadsheet application that runs on both Windows and Macintosh systems. It's part of the Microsoft Office suite, which also contains Word and PowerPoint, among other productivity tools.

Despite the fact that Excel is a Microsoft product, the first Macintosh version was launched in 1985. Excel was not made accessible for Windows until 1987, when Microsoft released Windows 3.0. Microsoft has continued to support the software on both platforms since then, providing upgrades every two years or so.

Figure 53. Excel Icon.

Source: Image by Pixabay.

IBM Lotus 1-2-3 (for Windows) and the AppleWorks spreadsheet application are two more prominent spreadsheet apps (for the Mac) Microsoft Excel, on the other hand, has dominated the spreadsheet industry for many years and remains the most popular spreadsheet program among businesses and consumers.

External Hard Drive

Internal hard drives are standard on nearly all personal computers. The operating system, programs, and other files are all stored on this drive. The internal hard drive provides enough disk space for most users to keep all of their programs and information. However, an external hard drive may be handy if the internal hard drive becomes full or if the user needs to back up the data on the internal hard drive.

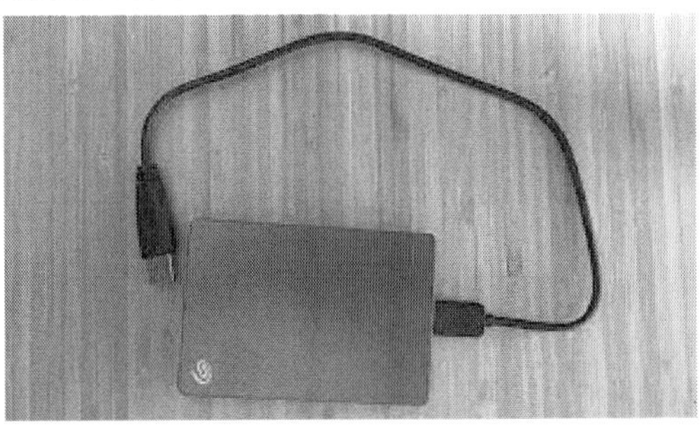

Figure 54. External Hard Drive.

Source: Image by Wikimedia.

USB hard drives commonly use the USB 2.0 interface because it supports data transfer rates of up to 480 Mbps. Even the most patient users would find the hard disk to be slow because USB 1.1 only permits transfers of up to 12 Mbps. Firewire drives enable data transmission rates of up to 400 and 800 Mbps, respectively, and can use either Firewire 400 or Firewire 800.

Audio and video editors are the most likely users to require external hard drives. This is due to the fact that high-quality media files can quickly fill even the most powerful hard disks. External hard drives, fortunately, can be daisy chained, which means they may be attached one after the other and utilized simultaneously.

This effectively doubles the amount of storage available. External hard drives might be beneficial for backing up your main hard drive even if you don't need extra storage. External hard drives provide excellent backups because they can hold an exact copy of another hard disk and can be kept in a secure location. Connecting the drive to the computer and transferring the relevant files from one device to another is all it takes to recover data or generate another backup.

While the majority of external hard drives come in bulky, protective casings, others are specifically built for the purpose of portability.

Extranet

You have a good comprehension of computer terminology if you can tell the difference between the Internet and an intranet. You might be in the upper tier if you know what an extranet is.

An extranet is a hybrid of the Internet and the intranet. It uses the Internet to connect an intranet, or internal network, to other users. The majority of extranets include a Web interface that may be accessed with a Web browser. Extranets often require authentication for users to access them since safe or secret information is frequently available within an intranet.

Unlike the Internet, "extranet" is not a proper noun and therefore should not be capitalized. A company's sensitive information with other firms or individuals frequently use extranets. For example, a supplier might use an extranet to give inventory data to a select group of clients while keeping the information private. A secure means of communication for the company and its clients, such as a support ticket system or a Web-based forum, may be included in the extranet.

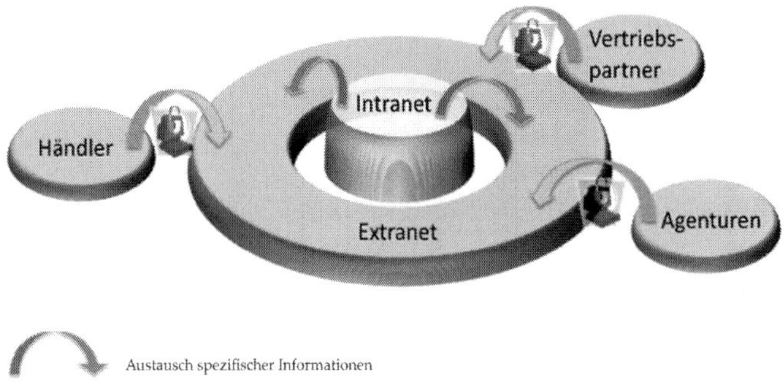

Figure 55. Extranet: A controlled private network.

Source: Image by Wikimedia.

F

Facebook

Facebook is a social networking website that was originally created for college students but is now available to anybody who is aged 13 and up. Facebook users may personalize their profiles by adding photographs, videos, and biographical information. Friends can look at each other's profiles and leave messages on their pages.

A "wall" is a section of a Facebook profile where friends can leave comments. Wall postings are essentially a public conversation because they are visible to all of the user's friends. As a result, it's generally not a good idea to write personal notes on your friends' walls. You can instead send a person a private message, which will appear in his or her inbox, similar to an email message.

Figure 56. Facebook icon.

Source: Image by Wikimedia.

Each Facebook user can customize their privacy settings, which are by default quite restrictive. For example, if you haven't added a specific individual as a friend, they won't be able to see your profile. You can, however, change your privacy settings to allow others in your network (such as your college or your neighbourhood) to see part or all of your profile. You can also build a "restricted profile," which allows you to hide some aspects of your profile from a specific group of users. You can add friends to your "restricted profile" list if you don't want them to be able to see your entire profile.

The option to add applications to your profile is another feature of Facebook that distinguishes it from MySpace. Small programs created exclusively for Facebook profiles are known as Facebook applications. SuperPoke (which enhances Facebook's "poke" function) and Fun Wall are two examples (which builds on the basic "wall" feature).

News feeds and weather forecasts are examples of informational applications. There are also hundreds of video game programs that allow users to play mini video games within their accounts, such as Jetman or Tetris. Friends can compete against each other or millions of other Facebook members because most game programs save top scores.

Facebook makes it simple for friends to stay in touch and for people to establish an online presence without having to create a website. Almost anyone can create a multimedia profile on Facebook because it is so simple to upload photos and videos. of course, whether you're a Facebook user or want to join one day, remember to be cautious about what you write or post on other people's accounts. After all, your personal information is only as public as you make it!

FAT32

The method Windows stores data on your hard disk is referred to by this unusual moniker. The "File Allocation Table," or "FAT," keeps track of all your files and assists the computer in finding them on the drive. Even if a file is fragmented (divided up into numerous locations on the disk), the file allocation table still can maintain track of it. FAT32 is a better version of the original FAT system since it uses more bits to identify each disk cluster. This makes it easy for the computer to identify data and allows for smaller clusters, which increases the hard disk's performance. FAT32 can store up to 2 terabytes of data on a hard drive. Unless you're a heavy power user, it should be enough of room for you.

Fiber Optic Cable

This is a cable that can convey light beams and is made up of ultra-thin glass or other transparent filaments. Because fiber-optic cables are made of light, data may be sent at the speed of light. Ones and zeros are conveyed via the connection using a laser transmitter that converts frequency signals into light pulses. The light signals are translated back into data that can be read by a computer at the receiving end of the transmission.

Fiber optics are less vulnerable to noise and interference than other data-transfer mediums such as copper wires or telephone lines since they are based exclusively on light beams. However, because the cables are delicate and are usually installed underground, installation is laborious and costly. Although some fibre-optic cables are located above ground, if they break, they must frequently be replaced in their entirety, which is costly. While copper wires can be spliced and repaired as many times as needed, glass fibre-optic cables are far more difficult to repair.

Figure 57. Fiber Optic Cable.

Source: Image by Wikimedia.

FIFO

It stands for First In, First Out. FIFO is a data processing and retrieval mechanism. The first items input are the first items removed in a FIFO system. To put it another way, the items are deleted in the same order that they were added.

To provide a real-life example, consider a vending machine with things loaded from the back. When a person chooses a Milky Way bar from row E5, the machine produces the candy bar that is closest to the front. The next Milky Way in line takes its place at the front of the line. As a result, the candy bars are dispensed in the order they were placed in the machine using the FIFO method.

When pulling data from an array or buffer, computers frequently use the FIFO system. FIFO approach (first in, first out) is utilized when the data that was entered first into the buffer is also the first one to be removed. LIFO approach (last in, first out) is the opposite of FIFO, in which the most recent data in is the first to be removed.

File

A file is a collection of data kept in a single unit with a unique name. It can be a paper, a photograph, an audio or video stream, a data library, an application, or any other type of data collection. A brief description of each file format is provided below.

Text files, such as Word documents, RTF (Rich Text Format) documents, PDFs, Web pages, and others, are examples of documents. JPEGs, GIFs, BMPs, and layered image files, such as Photoshop documents, are examples of pictures (PSDs). MP3s, AACs, WAVs, AIFs, and other audio files are examples. Video files can be encoded in a variety of formats, including MPEG, MOV, WMV, and DV, to mention a few.

A library file is a collection of data that is used by a program or the operating system. Plug-ins, components, scripts, and a variety of other items fall under this category. A program, or executable file, is what an application is. Both software and files, such as Microsoft Internet Explorer and Apple iTunes, exist.

Open, save, delete, and move files to various folders are all options. They can also be downloaded from the Internet or transmitted over network connections. The type of a file can be established by looking at the icon or reading the file extension. If a file type relates to a specific application, double-clicking it will usually open it in that software

File extension

"Common file extensions include ."txt" for text files and ."jpg" for JPEG picture files," as an example.

After the dot, the file extension is the last part of the filename (e.g., ."pdf"). The most common file extension is three characters long, although it can also be shorter or longer. The file type of each file is defined by its extension. This information is used by the operating system to determine which application will open the file and which icon will be assigned to it.

To see a comprehensive list of file extensions and their associated file types, go to the Sharpened File Extensions Reference.

Figure 58. File extensions.

Source: Image by Pixabay.

File Format

"PNG files store image data in a compressed binary file format," for example. The way data is saved in a file is described by the file format. It specifies the file's data structure (how the data is organized) as well as the type of data it contains. Some file formats are "open formats," which means that they are freely available to all software developers. Other file types are proprietary or "closed formats," which can only be opened by specified programs.

Text and binary data are the two types of data that can be stored in a file. Text-based file formats, such as XML and HTML, save data as plain text, allowing the contents of the file to be examined in a text editor. Binary files, on the other hand, can only be opened by a program that understands their format. While certain binary files can be opened using a text editor, the majority of the contents will be distorted and unreadable.

File type

A file type is a specific sort of file, as the name implies. Some file formats are specific to a single application, whilst others are universal and can be opened by a variety of programs. A Microsoft PowerPoint presentation (.PPTX extension) is an example of a proprietary file type that can only be accessed by Microsoft PowerPoint. Microsoft PowerPoint is the only program that can open it. Rich text documents (.RTF files) are a universal file type that can be opened by almost any text editor.

The extension of a file can often be used to determine its kind. A.DOCX file, for example, is a Microsoft Word document. The file extension is ."DOCX," and the file type is "Microsoft Word document." The application name is usually included in proprietary file formats, but not in universal file types. Because it is not associated with a specific application, a.JPG file is simply referred to as a "JPEG image."

Important: While the terms "file type" and "file format" are often used interchangeably, a file format technically refers to the way data is stored within a file.

Filename

A filename is a text string that is used to identify a particular file. Every file on a computer's hard drive has a filename that aids in locating the file within a folder. As a result, each file within a folder must have a unique name, although files in separate folders can have the same name. Letters, numerals, and other characters may appear in filenames.

Certain characters cannot be used depending on the operating system because they contradict with the operating system's operators or other syntax. The maximum number of characters a filename can have varies depending on the operating system. Newer operating systems allow filenames to be as long as 256 characters, whereas older operating systems only allowed for 8 or 16 characters. of course, 16 characters is usually plenty for most practical needs.

A file extension, which specifies the file type, is commonly included in filenames. Because it is appended to the filename after a dot or period, the file extension is sometimes known as the "filename suffix." A Microsoft Word document might be called "document1.doc," for example. While the filename in the above example is "document1" and the extension is "doc," it is also allowed to refer to the filename as "document1.doc." In rare circumstances, the filename may even refer to the directory in which the file is stored, i.e. ("\My Documents\School Papers\document1.doc").

By clicking on the file's icon or filename, waiting a second, and then clicking on the filename again, you can name the file. The filename will become highlighted as long as the file is not locked, and you can input a new name for it. You can also name a file by selecting "Save As..." from the program's File menu the first time you save it.

By clicking on the file's icon or filename, waiting a second, and then clicking on the filename again, you can name the file. The filename will become highlighted as long as the file is not locked, and you can input a new name for it. You can also name a file by selecting "Save As..." from the program's File menu the first time you save it.

Firewall

The term "firewall" originally referred to fire-resistant barriers that were intended to keep fires from spreading from one room or structure to the next. They were usually composed of brick, steel, or another inflammable material that would successfully stop a fire from spreading. Firewalls do not entail fire or pyrotechnics, but they serve a comparable purpose in the IT world. A computer firewall restricts the amount of data that may pass through it and guards against unauthorized users damaging a networked server or client machine.

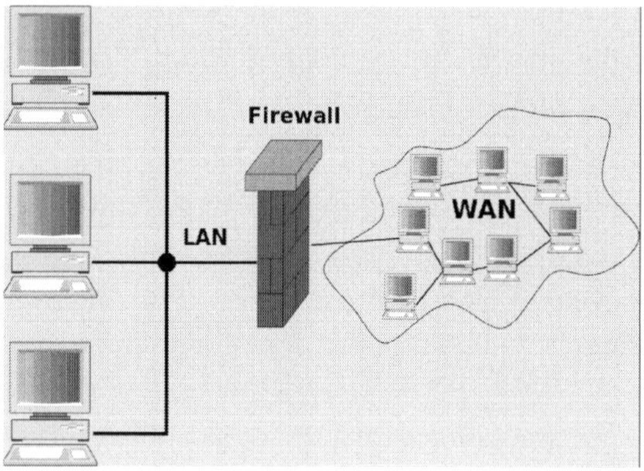

Figure 59. Firewall.

Source: Image by Wikimedia.

Hardware or software-based firewalls are available. A router is an example of a hardware device with an integrated firewall. Most routers can be configured to limit traffic from certain IP addresses or block requests based on other criteria. Firewalls are software applications that monitor and restrict external access to a computer or network. A network firewall permits only permitted Internet traffic into and out of the network.

Firmware

A firmware program or set of instructions is a software program or set of instructions that is installed on a hardware device. It contains the instructions for how the device interacts with the rest of the computer hardware. But how can you program software into hardware? That is an excellent question. Firmware is often stored in a hardware device's flash ROM. While ROM is "read-only memory," flash ROM is a sort of flash memory that can be erased and overwritten.

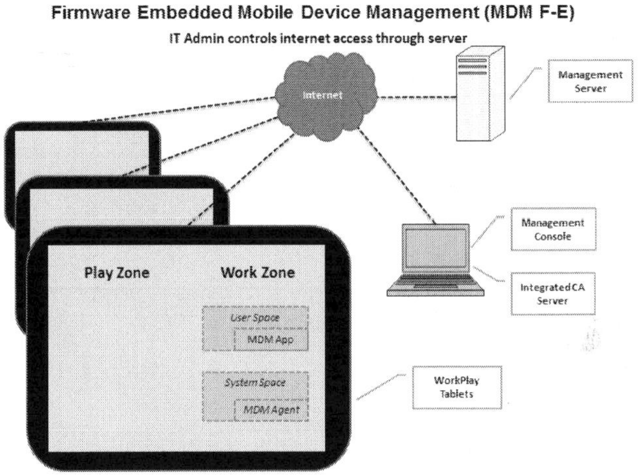

Figure 60. Firmware.

Source: Image by Wikimedia.

Firmware is considered "semi-permanent" since it does not change unless it is updated by a firmware updater. Certain equipment, including as hard drives and video cards, may require firmware updates in order to run with a new operating system.

Certain equipment, including as hard drives and video cards, may require firmware updates in order to run with a new operating system. Firmware upgrades for CD and DVD drives are frequently made available, allowing the devices to read faster media. Manufacturers occasionally provide firmware updates that simply improve the efficiency of their equipment.

Firmware updates are frequently found in the "Support" or "Downloads" sections of a manufacturer's website. Although it is not always required to keep your firmware up to date, it is still a good idea. Just make sure that once you start a firmware updater, you finish it because most gadgets won't work unless their firmware is recognized.

Flash

Flash is a multimedia programming language that may be used to create animations and interactive websites. Flash is frequently used by web developers to create dynamic content that is not possible to create using HTML or other scripting languages. Splash screen animations, YouTube videos, and interactive music playlists are just a few examples.

The Flash plug-in must be installed in order to view Flash content in your browser. Because Flash is currently used on so many websites, Adobe also offers a free download of the standalone Flash Player.

Flash Drive

Jump drives, thumb drives, pen drives, and USB keychain drives are all names for flash devices. Whatever name you give them, they're all referring to the same thing: a compact data storage device with flash memory and a built-in USB port. Flash drives usually have a length of two to three inches and a width of less than an inch. They may have the size and form of a thumb or a little pen (which is where the names "thumb drive" and "pen drive" come from) Flash drives are also extremely thin, with a depth of less than a centimeter in many cases.

They are incredibly portable due to their compact size and may easily fit in a pocket or on a keychain (hence the name "keychain drive"). Modern flash drives can hold many gigabytes of data, but early flash drives could only store a few megabytes. Flash drives have mostly superseded prior portable data storage mediums such as floppy disks and removable hard disks such as Zip disks due to their tiny size and enormous storage capabilities. Flash drives do not require the usage of a dedicated disk drive because they have a built-in USB connection. They can instead be used on any computer having a USB port, which almost all recent PCs have.

Flash Memory

Flash memory refers to an electrically erasable programmable read-only memory (EEPROM). The term derives from the memory's design, which allows a chunk of memory cells to be erased in a single operation or "flash." The BIOS settings in a computer's ROM are a frequent use of flash memory. When the BIOS needs to be updated, the flash memory can be written in blocks rather than bytes, which makes updating the BIOS much easier. For the same reason, most modems use flash memory.

Though flash memory was first employed in computers, it has since spread to a variety of different applications. Digital cameras, cellular phones, networking devices, and PC cards all use flash memory cards. Though the memory's read/write performance isn't lightning fast, having a little card to carry around rather than a large hard drive is convenient.

Floating Point

Floating point numbers, as the name implies, are numbers with floating decimal points. The numbers 5.5, 0.001, and -2,345.6789, for example, are all floating-point numbers. Integers are numbers that do not have any decimal places.

Real numbers with fractions are recognized by computers as floating-point numbers. A "floating point calculation" is a calculation that includes a floating-point number. Previously, these computations were done by a separate floating-point unit (FPU) but nowadays the FPU is usually incorporated into the computer's CPU.

Floppy Disk

Even though they do not appear to be "floppy," floppy disks have an intriguing moniker. When you remove the disk out of the protective covering, you'll notice that it's quite flexible. It's coated with iron oxide and works similarly to a hard disk in that it saves data magnetically.

In 1969, the first floppy disks were manufactured, the same year that the Internet was launched. These disks were 8 inches in diameter and read-only, similar to a CD-ROM, meaning the user could not write data to them. The earliest 8-inch disks could only store 80KB of data, but later versions could store up to 800KB.

In 1978 a 5.25-inch disk with 360KB of storage capacity was launched. Later iterations of the 5.25-inch floppy disk could store 1.2 megabytes. Early desktop computers, such as the Apple IIe, utilized these disks. In 1987, the 3.5-inch HD (high density) floppy disk was introduced, with a capacity of 1.44 megabytes after formatting. These disks were a little more robust and portable than the 5.25-inch disks. The 3.5-inch floppy disk became the main method of distributing commercial software titles and keeping up personal data for the next decade.

In the late 1990s, CD-ROMs began to supplant floppy disks as the preferred method of software distribution. Consumers began switching to recordable CDs for data backup a few years later. The first mainstream computer without a floppy disk drive was Apple's iMac, which was released in 1998. Many PC makers soon followed suit, but it took several years.

The majority of software is now provided on CDs and DVDs, and the majority of consumers back up their data on recordable CDs or USB flash drives. Floppy disks are finally being phased out, which is a good thing because they are infamous for causing data loss. Even Nevertheless, because it was the data storage medium with which many people grew up, the floppy disk will always hold a special place in the hearts and minds of veteran computer users.

Freeware

Freeware, like shareware, is software that you can download, share, and distribute without having to pay a fee. Freeware, on the other hand, has the advantage of never having to pay for it. It's completely free, with no 30-day limit, demo versions, or blocked functionality. Freeware is widely used to distribute minor program updates and small games. Even if freeware is free, it is still copyrighted, which means that others cannot sell it as their own.

Frozen

While the state of Minnesota is "frozen" from November to March, it also refers to a computer that is unresponsive. A computer is frozen when it does not respond to any user input. When a computer system freezes, or "locks up," the screen remains unchanged regardless of whatever buttons on the mouse or keyboard are pressed.

When you move the mouse, the cursor will not move, indicating that your computer has frozen. A computer usually freezes as a result of a software bug that causes the operating system to "hang." This can happen for a variety of reasons, including a memory leak, an infinite calculation, or something else entirely. A hardware failure, such as a defective RAM chip or a processor error, can also cause a computer to freeze.

A software crash is frequently caused by a software programming fault or unrecognized input, as computers are not designed to freeze. Modern operating systems, such as Mac OS X, are structured in such a way that if one program fails, it does not affect other programs and the computer does not freeze. If your computer does freeze, you'll have to restart it to get it to work again. By pressing and holding the power button for many seconds, you can usually force your computer to shut down. Also, because most computer freezes happen out of the blue, it's a good idea to save your work often!

FSB

The FSB, which stands for "frontside bus," connects the computer's processor to the system memory (RAM) and other motherboard components. The system chipset, AGP card, PCI devices, and other peripherals are among these components. Like the processor, the frontside bus's speed is measured in Megahertz or Gigahertz. The FSB speed is often a ratio of the processor speed because most computers' processors are faster than their system buses. For example, a Pentium 4 processor with a 2.4 GHz clock speed may only have a 400 MHz FSB. The ratio of CPU to FSB would be 6:1. A Power Mac G5 with a 2.0 GHz processor, on the other hand, has a 1.0 GHz frontside bus. As a result, the CPU-to-FSB ratio is 2:1.

The processor can perform more efficiently if the ratio is less. As a result, quicker front-end bus speeds result in improved overall performance. When the CPU to FSB ratio is high, the processor must frequently wait for data to be passed out across the system bus before it can process fresh data. As a result, the FSB might be a performance bottleneck in a computer. So, when looking for a fast computer, don't just look at the processor speed; also look at the frontside bus speed.

FTP

It stands for File Transfer Protocol. It is a widespread means of moving files from one computer to another via the Internet. "Fetch" for Mac and "WS FTP" for Windows are two popular FTP apps. FTP servers can also be accessed using a Web browser such as Netscape or Internet Explorer. To do so, type the URL into your browser of the server into the location field of the browser.

For example, "ftp:/ftp.servername.com/" will show you all of the FTP server's directories, "ftp:/ftp.servername.com/directory/" will show you all of the files in that directory, and "ftp:/ftp.servername.com/directory/filename" will download the file to your computer. Many FTP servers are "anonymous FTP" servers, meaning you can access them using the user's name "anonymous" and your e-mail address as the password. To access the files on other FTP servers, you'll need to create a unique login.

G

Gateway

A gateway is a piece of hardware or software that functions as a link between two networks, allowing data to flow across many computers. When you send an e-mail to a friend or log in to a web site, there is a gateway that permits the connection to happen. Many smaller connections to other servers will be made along the way while you connect to a web site. A number of gateways are employed in these situations. Gateway is also the name of a popular direct-order PC maker in an entirely other narrative.

Gibibyte

A gibibyte is a data storage unit with the value of 2 to the 30th power, or 1,073,741,824 bytes. A gibibyte is specified as exactly 1,073,741,824 bytes, while a gigabyte is approximated to be 109 or 1,000,000,000 bytes. This eliminates the ambiguity surrounding the number of gigabytes. A gibibyte is 1,024 mebibytes in size and comes before the tebibyte.

GIF

You don't have to recall that the initials "GIF" stands for "Graphics Interchange Format." A GIF is a compressed picture file format, which you should be aware of. GIF pictures are compressed using an algorithm created by CompuServe (which is why you see the term "CompuServe GIF" in Photoshop, for those of you that care). GIFs use an indexed color palette, which has a maximum of 256 colors.

This significantly reduces the size of their files. These compressed picture files can be sent over a network or the Internet rapidly, which is why you'll see them on a lot of web pages. GIF files are ideal for small icons and animated graphics, but they lack the color gamut needed for high-resolution photos.

Gigabyte

A gigabyte is 1,073,741,824 bytes, or 2 to the 30th power. It's around one billion (1,000,000,000) bytes, or 10 to the 9th power. A gigabyte is 1,024 megabytes in size, and it comes before the terabyte. The size of a hard disk is usually expressed in gigabytes, such as a 160GB or 250GB drive. In speech, the term gigabyte is frequently shortened as "gig." For example, if you have a 250GB hard drive, you could say, "I have 250 gigs of disk space." The prefix "giga" comes from the Greek word "gigas," meaning giant.

Gigahertz

1,000 megahertz (MHz) or 1,000,000,000 Hz equals one gigahertz. It's a standard way of determining how fast a computer can process information. For a long time, computer processor speeds were measured in megahertz, but as personal computers passed the 1,000-megahertz threshold around the year 2000, gigahertz became the official measurement unit. After all, "2.4 Gigahertz" is easier to say than "2,400 Megahertz."

While gigahertz is most usually used to measure processor speed, it can also be used to measure the speed of other computer components like RAM and backside cache. The entire performance of the computer is influenced by the speed of these components, as well as other sections of the computer. When comparing computers, keep in mind that the number of gigahertz isn't the only factor to consider.

GIGO

It stands for Garbage In, Garbage Out. It indicates that if erroneous data is entered into a computer program, the output will be erroneous as well. So, if a program asked you to enter a letter of the alphabet and you chose "3.14159" as a joke, there's a good probability the results you'd get would be jumbled up, or "junk." Programmers must take great efforts to prevent GIGO problems because computer users aren't always clever enough to submit proper data.

GNU

GNU (a recursive acronym for "GNU's Not Unix") is a Unix-like operating system that is available in several distributions. The GNU/Linux operating systems are based on the Linux kernel and are collectively known as GNU/Linux. gNewSense, Trisquel, and Venenux are some popular examples. For example – "The developer installed a GNU operating system on his secondary computer."

Without requiring a commercial license, all GNU systems and programs are distributed as free software. Instead, they're released under the GNU General Public License (or GPL), which specifies that the software can be freely used and distributed as long as other users have the same rights. While the GNU General Public License (GPL) While the GNU General Public License was created with GNU software in mind, it is currently utilized by a wide range of other free software programs.

Golden master

"The newest release candidate of the software is intended to be the golden master," for example. The golden master stage of the software development process is the last stage. It refers to a software package's "Release to Manufacturing" or RTM version, which is submitted to a manufacturing facility for reproduction. When software achieves the golden master (GM) version, it has completed development and will shortly be released to the public.

Goodput

"Network congestion may cause the goodput to drop relative to the throughput," for example. The term "goodput" sounds like a golf term for making a 20-foot birdie putt. Goodput is similar to throughput in the computer industry, which quantifies the average data transfer speed via a communications channel.

The amount of data transferred is divided by the time it takes to transport the data to compute throughput. Packet headers, acknowledgements of packet receipt, and retransmitted data are all examples of this. The original data is split by the transmission time to determine the goodput.

For example, during the data transmission process, a 5-megabyte file may require 300 kilobytes of header information and acknowledgements. As a result, the throughput is around 5.3 megabytes divided by the transfer time. The initial 5 megabytes divided by the transmission time would be the goodput. As a result, goodput is always less than or equal to the measurement of throughput.

Google

Google is the most widely used search engine on the planet. Larry Page and Sergey Brin, two Ph.D. students at Stanford University, started it as a search project in 1996. They devised a search engine algorithm that rated Web pages based on how many other Web pages they linked to as well as their content and keywords.

Google's Web search market share grew rapidly as a result of this method, which delivered more helpful results than other search engines. In September 2001, the Google ranking algorithm was renamed "PageRank" and patented. Google rose to become the world's most popular search engine in a short period of time.

Google's mission statement is to "organize the world's information and make it widely accessible and valuable," according to the company's website. While Google's primary tool for helping consumers find information is still its Web search, the corporation also provides a number of additional services. Here are a few examples:

1. Froogle is a pricing comparison website.
2. Image Search – look for photos on the internet.
3. Google Groups – discussion forums on the internet
4. Google Answers – based on a bidding structure, Google Answers provides answers to inquiries.
5. Google Maps – directions and maps
6. Google Toolbar Is a search tool that can be downloaded.
7. Blogger is a free blogging platform.
8. Gmail – Web-based e-mail with a storage capacity of several gigabytes
9. AdWords – Advertisers' advertising services
10. AdSense – A service that provides advertising to web publishers.

Google has become such a well-known search engine that the phrase "Google" is now frequently used as a verb to mean "search." For example, if you want to learn more about someone, you can use Google's search engine to Google that person.

Google Drive

Google Drive is a cloud-based file storage and sharing service provided by Google. The service, which began on April 24, 2012, offers 5 GB of free storage. A monthly fee can be paid for additional storage. Google Drive's purpose is to give you a central location to store your files online that you can access from anywhere.

You may also access your Google Drive account from a variety of devices, as the program is available for Windows, Mac OS X, Android, and iOS. A web-based interface is also available, allowing you to organize your files and search for documents by filename or content. Google Drive offers capabilities for sharing files and collaborating on projects with other users over the Internet in addition to online file storage. Instead of sending huge attachments via email, you can send one or more users URLs to files stored in your Google Drive. You can also create and edit documents online using the web-based Google Docs software.

Google Drive lets you view over 30 different file kinds right in your browser. Google's proprietary formats, as well as other common file types like Adobe Photoshop and Illustrator documents, are among them. Visit FileInfo.com's Google Drive File Types article for additional information on Google Drive's proprietary file types.

GPS

The Global Positioning System (GPS) is a satellite navigation system that determines ground position and velocity (location, speed, and direction). Despite the fact that it was developed and first used by the United States military, all around the world, GPS is now available to the general population. A lot of luxury cars now have GPS navigation systems, which include an LCD map that displays the driver exactly where he is in the world. Advanced automobile GPS units can speak directions to a specific location and inform the driver when to turn.

Figure 61. Global positioning system.

Source: Image by Flickr.

GPU

Stands for "Graphics Processing Unit" It's a single-chip processor, just like the CPU (Central Processing Unit). The GPU, on the other hand, is largely utilized to compute 3D functions. Lighting effects, object modifications, and 3D motion are all examples of this. Because these calculations are heavy on the CPU, the GPU can aid the computer's performance. NVIDIA, Inc. was the first company to develop the GPU.

Its GeForce 256 GPU has around 22 million transistors and can compute 10 million polygons per second. In comparison, the Pentium III chip has 9 million transistors. Wow, that's a significant amount of computing power. The processor also comes in a workstation version called the Quadro, which is targeted for CAD applications. This processor can perform up to 17 million polygons per second and execute over 200 billion operations per second.

Graphics

Images shown on a computer screen are known as computer graphics. They may be two-dimensional or three-dimensional. Raster and vector graphics are two types of two-dimensional graphics. For example modern video games have 3D graphics that can appear incredibly realistic.

Raster graphics is the most common type of computer graphics. They are used to create icons, pictures, and other simple images. Drawings, logos, and other scalable objects are created using vector graphics. Polygons make up 3D visuals, which can be developed using CAD and 3D modelling software. Video games and 3D animations are the most popular places where they can be found.

GUI

It's pronounced "gooey" and stands for "Graphical User Interface." It refers to a computer's graphical interface, which lets users to click and move items with a mouse rather than typing commands at a command line. Windows and Mac OS, two of the most popular operating systems, are GUI-based. Apple introduced the graphical user interface to the world with the Macintosh in 1984. However, the concept was inspired by a Xerox user interface from years ago.

"Globally Unique Identifier" is the abbreviation for "Globally Unique Identifier." A GUID is a 128-bit (16-byte) number that software programs use to uniquely identify a data object's location. Streaming media files,

Windows registry entries, database keys, and numerous file kinds are all examples of data that contain GUIDs. GUIDs are commonly written in hexadecimal format with 32 digits and look like this:

{12345678–9012–3456–7890–123456789012}

The GUID Partition Table is similarly based on globally unique IDs (GPT). Intel suggested this hard disk partitioning mechanism as part of the Extensible Firmware Interface. Windows PCs and Intel-based Macintosh machines both use it. GUIDs are used by GPT to identify the various partitions on a hard drive. The boot partition, file system partition, and data partition are just a few examples. Each operating system that supports the GPT partitioning technique labels each partition with a unique GUID.

H

Hacker

While this phrase used to indicate to a skilled or knowledgeable coder, it is now more generally used to describe someone who can get unauthorized access to other computers. A hacker can "hack" his or her way through a computer systems or network's security levels.

This can be as easy as figuring out someone else's password or as difficult as building a custom program to defeat the security software on another computer. Hackers are the reason why software companies provide "security upgrades" on their products on a regular basis. While the average person is unlikely to be "hacked," some huge firms and organizations face multiple hacking attempts every day.

Figure 62. Hacker: Exploits a computer network or system.

Source: Image by Pixabay.

Halftone

Instead of a continuous tone, a halftone image is made up of a sequence of dots. These dots can come in a variety of sizes, colors, and occasionally even forms. Darker, more dense regions of the image are represented by larger dots, whereas lighter areas are represented by smaller dots.

Newspapers and periodicals utilize halftone graphics because they are a more efficient way to publish photographs. A halftone image uses substantially less ink to print than a continuous tone image because it is made up of discrete dots. The dots look as a continuous image to the human eye as long as the image resolution (measured in LPI) is high enough. You should be able to see the dots that make up the halftone image if you look closely at the photographs printed in a newspaper.

Handle

The name you use in online chat is sometimes referred to as your screen name, or handle. So, if you wanted to be known as MooCow123, that'd be your handle. So, if you wanted to be known as MooCow123, that'd be your handle. Handles are useful because they let you express yourself without revealing your true identity. The phrase comes from CB radio, where users use their "handles" to refer to each other.

Hard Copy

A printed document is referred to as a hard copy. It might be a text file, a photograph, a drawing, or anything else that can be printed. Instead of sending a business memo by e-mail, it could be sent as a hard copy, or a real piece of paper carrying the memo.

When you produce a document on a computer, you usually save it as a file on the hard drive. A soft copy is a term used to describe this type of document. On a computer, the file can be easily opened and changed, but it may also be removed. As a result, printing a hard copy of a document is occasionally done to create a physical backup of the document.

Hard Disk

The information is often written to your hard disk when you save files or install programs on your computer. The hard disk is a spinning disk with magnetic platters that record and store data. Information recorded to the hard disk remains intact after you turn your computer off since it is stored magnetically. The hard drive differs from RAM, or memory, which is reset when the computer's power is turned off.

The hard disk, which reads and writes data to the disk, is located inside the hard drive. Data is also sent back and forth between the CPU and the disk via the hard drive. When you save data to your hard disk, thousands, if not millions, of ones and zeros must be written to the hard disk. It's a fascinating process to contemplate, but it's also an excellent reason to preserve a backup of your data.

Hard Drive

All of your data is stored on the hard disk. It houses the hard disk, which contains all of your files and folders. A normal hard drive is around the size of your hand, but it can store over 100 GB of data. The information is kept on a stack of disks that are housed in a solid casing.

These disks spin extremely quickly (usually at 5400 or 7200 RPM), allowing data to be accessible instantly from any location on the drive. Because the data is magnetically recorded on the hard disk, it remains on the drive even when the power supply is turned off. To increase security, some hard tokens are used in conjunction with other security measures. To get access to a protected system, for example, a username and password or a fingerprint scan may be required in addition to the hard token.

Another sort of hard token is a hard tag, which is used as a security device on items like clothing and gadgets. You've probably seen these at a number of retail outlets where they're employed to prevent theft. The security system at the doors beeps if someone tries to escape the facility with items before the hard tag has been deactivated. of course, even though you paid for the product, if the cashier forgets to disable the hard tag, you will hear the annoying beeping sound as you left.

Hardware

The actual components of a computer and related devices are referred to as computer hardware. Motherboards, hard drives, and RAM are examples of internal hardware devices. Monitors, keyboards, mouse, printers, and scanners are examples of external hardware.

Components refer to the internal hardware components of a computer, whereas peripherals refer to external hardware devices. They all fall under the category of computer hardware when taken together. Software, on the other hand, is made up of the programs and applications that run on a computer. Because software operates on computer hardware, it is common for software programs to include system requirements that specify the minimal hardware that must be installed in order for the software to function.

HDD

"Hard disk drive" is abbreviated as HDD. A hard disk drive (HDD) is a data storage device. A hard disk drive, unlike RAM, stores data magnetically and does not require power to retain its state. As a result, even if the power supply is switched off or disconnected, the data is retained.

Your files are stored on either an HDD or an SSD in most PCs. If your computer has an internal hard disk drive, for example, when you save a file, it is written to the hard disk drive. In addition, the HDD saves pre-installed applications as well as any additional programs you install on your computer. Most users will never run out of disk space because contemporary HDDs have storage capabilities of more than one terabyte. You can, however, connect an external HDD (also known as an external hard drive) for additional storage or backup purposes if necessary.

Important: The terms "hard disk drive," "hard drive," and "hard disk" are all interchangeable. An HDD and a hard drive are the same thing in terms of technology; however, a hard disk refers to the magnetic platter inside the drive enclosure.

HDMI

Stands for "High-Definition Multimedia Interface." HDMI is a digital interface that allows audio and visual data to be transmitted via a single cable. Most HDTVs and associated components, such as DVD and Blu-ray players, cable boxes, and video gaming systems, support it.

Other forms of A/V connections require separate wires for audio and video data, whereas HDMI combines the two streams, substantially reducing cable clutter. A component cable connection, for example, necessitates three video cables and two audio connections, for a total of five wires. Only one HDMI cable is required to convey the same data.

HDMI cables are less susceptible to interference and signal noise than analog connections because they are digital. Furthermore, because most components, such as DVD players and digital cable boxes, handle data digitally, HDMI avoids the analog to digital conversion required by traditional interfaces. As a result, when compared to other types of connections, HDMI frequently offers the best picture and sound quality.

HDMI cables are often more expensive than analog cables, owing to the higher manufacturing costs. It's worth noting, though, that HDMI eliminates the need for separate audio and video lines. Furthermore, for people who dislike dealing with complicated wires and connectors, the single all-purpose connection may be worth the price difference. Just remember to double-check that the devices you're connecting have HDMI ports before purchasing an HDMI cable.

HDTV

HDTV is a high-definition video standard that was created to replace other video formats such as SDTV (standard definition television). While the video quality of HDTV is one of the most visible differences from SDTV, it also features a number of other significant enhancements. First and foremost, HDTV is a digital signal. HDTV is always digital, unlike standard NTSC broadcasts, which use an analog feed. Electrical currents and magnetic fields generate analog interference, which is eliminated.

Second, HDTV has a different aspect ratio than standard definition television. HDTV employs a 16:9 ratio, whereas prior broadcasts used a 4:3 ratio (4 units wide for every 3 units tall). The image appears more lifelike because of the larger aspect ratio, which more closely resembles how people experience the environment. This aspect ratio is also preferable for watching widescreen movies, which are also shot in widescreen for the same reason.

High-definition television, as the name implies, has a far better resolution than standard definition TV. While a normal analog broadcast in the United States has a maximum resolution of 525 horizontal lines, an HDTV signal can handle up to 1080. 1080i (interlaced), 720p, and 1080p are the three formats used by HDTV (progressive). The higher resolution of HDTV results in visuals that are finer, with more detail and color than prior formats. HDTV also offers a higher-quality digital audio stream than SDTV, with up to six audio channels available, as opposed to the two previously allowed.

You'll need an HDTV-compatible television and a way to receive an HD signal to watch HDTV. HDTVs are available in both 16:9 and 4:3 aspect ratios (for backwards compatibility). Some HDTVs come with HDTV tuners for receiving over-the-air broadcasts, while others require the purchase of a separate receiver. Most cable and satellite TV companies, fortunately, include HDTV-compatible equipment as part of their digital service package.

Heat Sink

Every second, a computer's CPU may do millions of calculations. The CPU generates heat as it continues to function at a high rate. The processor may overheat and finally harm itself if this heat is not controlled. Fortunately, CPUs have a heat sink that dissipates the processor's heat and prevents it from overheating. The heat sink is constructed of metal, such as zinc or copper alloy, and is connected to the CPU by a thermal substance that draws heat away from the processor and towards the heat sink. Heat sinks can be as little as just covering the processor to as large as several times the processor's size if the CPU necessitates it.

Most heat sinks also include "fins," which are thin metal slices that attach to the heat sink's base. These extra pieces of metal help to dissipate heat by distributing it across a broader area. A fan is frequently used to cool the air around the heat sink, preventing it from becoming excessively hot.

A heat sink and fan (HSF) combo is the name for this setup. Heat sinks are used in almost all computer CPUs, and they have also become prevalent in visual card processors, or GPUs.

Hertz

The standard unit of measurement for frequency is the Hertz (abbreviated as Hz). A hertz equals one cycle per second since frequency is measured in cycles per second.

Sound, light, and radio waves are all measured in Hertz. Sound waves between 20 and 20,000 Hz, for example, are detectable by the typical human ear. Sound waves with a frequency of less than 20 Hz are referred to as "bass" frequencies because they have a low pitch. Above 5,000 Hz, sound waves have a frequency.

Hertz is a unit of measurement that can be used to measure wave frequencies as well as the speed of computer processors. Each CPU, for example, has a fixed clock speed. This figure represents the number of instructions cycles the CPU can complete in a second. Clock speeds are often measured in megahertz or gigahertz since current processors may run millions or even billions of instructions per second.

Hexadecimal

A base-16 number system is hexadecimal. It's a different way of representing numbers than the base-10 system that we're used to. In base-10, people count in multiples of 10 before adding another digit. "8 – 9 – 10 – 11 – 12..." and "98 – 99 – 100 – 101 – 102..." are two examples. When the number 10 is reached, a new digit is added, then another digit is added to indicate 100.

(10x10). Each digit in the base-16, or hexadecimal, number system can have sixteen values instead of ten. A hexadecimal digit can have the following values:

A, B, C, D, E, F, G, H, I, J, K, L, M, N, O, P, Q, R, S, T, U, V, W, X,

As a result, in hexadecimal notation, the number 12 (in the standard base-10 format) would be written as "C." The sum of the numbers 18 (16+8) equals 24. In hexadecimal, 100 equals 64 (16x6 + 4) and 1000 equals 3E8 (256x3 + 16x14 + 8).

While computers use the base-2, or binary, system to analyze numbers, it is often more effective to visualize the numbers in hexadecimal format. This is due to the fact that four binary digits are represented by only one hexadecimal digit. Because a byte contains eight binary digits, only two hexadecimal digits are required to represent one byte.

HFS

It stands for "Hierarchical File System." The HFS file system is used to organize files on Macintosh hard drives. The hierarchical file system is used to establish a directory that can expand when new files and folders are added to the drive when a hard disk is formatted for a Macintosh computer. Windows machines do not recognize HFS-formatted drives because it is a Macintosh format. WIN32 or NTFS file systems are commonly used to format Windows hard disks.

Because HFS was not meant to handle huge hard disks, such as the 100GB+ hard disks that are prevalent today, Apple released Mac OS 8.1 with an improved file system dubbed HFS+, or HFS Extended. HFS+ enables for smaller clusters or block sizes, lowering the minimum file size required. On huge hard disks, this means that disk space may be used considerably more efficiently. By default, Mac OS X employs the HFS+ file system, which also allows journaling.

Hibernate

It is a good idea to put your computer to sleep (also known as standby mode) if you are not intending to use it for a few hours. It is normally advisable to turn off the computer if you are not intending to use it for more than a day. However, certain Windows computers have a feature called "Hibernate" that combines the benefits of sleep mode and turning down the machine.

When you choose Hibernate, the computer stores the current state of the system to the hard disk before shutting down. Instead of going through the standard boot procedure, the previously saved state is automatically loaded into RAM when the computer is restarted. The open windows and apps from your previous session display just as they did when the computer went into Hibernate mode. Hibernate mode is similar to Standby mode, except that the machine can be totally shut off. This is especially useful for laptop computers, which, if left in sleep mode, will gradually lose their battery charge.

Select "Turn Off Computer" from the bottom of the Windows Start Menu to put your computer to sleep. When the dialog box displays, hit, and hold the Shift key. "Standby" should be replaced with "Hibernate."

If a web page has five images, six "hits" will be registered when the page is loaded. This is due to the fact that the browser must send six requests to the Web server: one for the HTML of the page and five for the images.

Hit

A hit is a metric which is used in website analytics. "Hits" and "visits" are used interchangeably, at times, but they are two different things. A visit is recorded when a user visits a webpage. while, on the other hand, a hit is recorded for each resource which is downloaded from a web server. Thus, it is common for hits to outnumber visits, usually by a ratio of more than 10 to 1.

Most of webpages reference multiple files, such as images, JavaScript files and CSS documents. Each local file referenced in the HTML counts as a hit, together with the webpage itself. Hence, if a page consists of eight images, references two CSS documents, and links to three JavaScript files, it would record 14 hits each time a visitor views the page. As website hits can differ based on how many resources each page contains, the metric is hardly used for measuring the website traffic.

Horizontal Markets Software

A horizontal market is one that sells to multiple industries rather than just one. As a result, horizontal market software is software that may be applied to a variety of industries. Word processing and spreadsheet programs, for example, are horizontal market applications since they are employed by a wide range of enterprises and consumers. A business owner might use word processing software to send notes to his staff, while a student might use the same tool to complete a class paper. Developers of software for horizontal markets often have a large audience, but they also face a lot of competition. Vertical market software is the opposite of horizontal market software, as it is solely produced for a single industry.

Home Page

A website's beginning point, or first page, is this. This page normally has a table of contents and frequently outlines the site's aim. Apple.com's home page, for example, is http://www.apple.com/index.html. When you write in a simple URL, such as http://www.cnet.com, you are usually taken to the site's main page. Many people have a "personal home page," which is a different phrase for "home page."

Hyperlink

A hyperlink is a word, phrase, or image that can be clicked to take you to a different document or area within the current one. Hyperlinks can be found on practically every Web page, allowing users to go from one page to the next by clicking on them. Text hyperlinks are commonly blue and highlighted, but they aren't required. The arrow should transform to a little hand pointing at the link when you move the cursor over a hyperlink, whether it's text or an image. When you click it, it will open a new page or a section of the current page.

Hyperlinks, often known as "links," are commonly seen on Web pages, although they can also be found in other hypertext documents. Certain encyclopedias, glossaries, dictionaries, and other hyperlinked references fall under this category. The links function similarly to those found on the Internet, allowing the user to go from one page to the next. Hyperlinks, in a nutshell, allow individuals to peruse information at breakneck speed.

Hypermedia

Clicking text-based links that open new pages in a Web browser is how most people navigate the Internet. These hypertext connections, which are frequently blue and underlined, allow the user to jump from one page to the next. Hypermedia is a type of hypertext that allows you to link images, movies, and Flash animations to other content.

An image link is the most prevalent sort of hypermedia. On the Internet, photos and graphics are frequently linked to other pages. Clicking on a small "thumbnail" image, for example, may launch a larger version of the image in a new window. You may be directed to an advertiser's website by clicking on a promotional graphic. By integrating one or more links that appear during playback, Flash animations and movies can be turned into hyperlinks.

Moving the cursor over an image or video will reveal whether it is a hyperlink. The image or video is connected to another page if the cursor transforms to a little hand. When you click on the text, image, or video, it will take you to a new page on your browser. As a result, only click a hypertext or hypermedia link when you're ready to exit the current page. You can usually right-click the link and choose "Open Link in New Window" to open it in a new window.

Hypertext

Hypertext is text that contains hyperlinks to other sources of information. A user can swiftly go to different material by clicking on a link in a hypertext document. Though most people equate hypertext with Web sites, the technology has been present since the 1960s. Dictionary and encyclopedia software packages have long included hypertext in their definitions so that users can easily learn more about specific words or topics. Hypertext was also employed in Apple Computer's HyperCard application, which allowed users to create multi-linked databases. Today, hypertext reigns supreme on the Internet, with practically every page containing links to other pages and both text and graphics capable of linking to other content.

I

I/O

I/O is an abbreviation for "Input/Output," and it's pronounced "eye-oh." Computers are built on the premise that every input produces an output. If you input a sentence on your keyboard while using a word processor program, the text will show on the screen. The screen is an output device, while the keyboard is an input device. Another output device is a printer, which you can use to print the text. All I/O activities are handled by the computer's CPU, which sends the data it receives to the relevant path.

The path could lead to the video card, the hard drive, or the RAM, to mention a few possibilities. Because they connect input and output devices to the computer, the ports on the outside of a computer are commonly referred to as "I/O ports." I/O is a term used by software developers to explain how a program will behave depending on what a user enters. For example, in the game "Super Jumper Man," if the user clicks the space bar key, the character on the screen will jump. Multiply that by thousands of other user input scenarios, and you've got yourself a computer game.

I/O Address

Each I/O (Input/Output) device attached to your computer has its own I/O (Input/Output) address. Every I/O port on your computer, including USB, Firewire, Ethernet, VGA, and DVI connectors, as well as any other ports your machine may have, is given one of these addresses.

Each port has its own unique address, which helps your computer to recognize and locate devices connected to it. The computer can identify a keyboard, mouse, monitor, printer, or any other device by its I/O address. I/O addresses do not consume any system memory or RAM because they are managed by the computer's motherboard.

IBM Compatible

Apple and IBM computers dominated the personal computer industry in the early 1980s. Apple's computers used a proprietary operating system designed by Apple, whereas IBM's computers used PC-DOS. IBM chose to license the DOS operating system to other manufacturers as the demand for personal computers grew. These businesses began manufacturing personal computers known as PC clones or IBM compatibles.

Supplies rose as more manufacturers began to produce PCs, and prices began to fall. As a result, more individuals could afford personal computers, and IBM compatibles began to dominate the market. It wasn't long until the quantity of computers sold directly by IBM was surpassed by the new manufacturers. When Microsoft released the Windows 95 operating system in 1995, sales of IBM compatibles skyrocketed once more. By that time, however, the name "IBM compatible" had practically become obsolete, as most PCs ran Microsoft Windows as their primary operating system. Furthermore, PC manufacturers had been constructing their own computers for many years, and IBM's own PCs and IBM compatibles had little similarities.

IBM ceased production of personal computers in 2005. The company that pioneered the personal computer revolution is no longer in business. As a result, the term "IBM compatible" is a little out of date, yet it can still be used to denote computers that run on Windows.

ICS

"Internet Connection Sharing" is what it stands for. Multiple PCs can connect to the Internet using the same Internet connection and IP address thanks to ICS. A router, for example, can connect multiple computers in a household to the same cable or DSL modem. Every computer linked to the router is connected to the Internet as long as the router is connected to the modem. The machines can share the same IP address thanks to network address translation (NAT).

Software can also be used to perform ICS. Internet connection sharing is available in Windows 98 and later, as well as Mac OS X. This allows you to change the network settings on a single computer, thereby turning it into a gateway. The Internet connection of that computer can then be used by other computers on the same network. Users of Windows may achieve the same outcome with apps like WinGate and WinProxy. While software can be used to share an Internet connection, using hardware (such as a router) for ICS is the simplest and most convenient option.

ICT

ICT is short for "Information and Communication Technologies." It's comparable to IT (Information Technology), but it's more focused on telecommunications technologies like the Internet, cell phone networks, and satellite technology. Users all around the world can now speak with one other in real time on a daily basis thanks to modern types of ICT. Examples include instant messaging, video-conferencing, online multiplayer gaming, and social networking websites. For example: "Thanks to ICT, users can access information from all over the globe."

IDE

IDE can refer to either "Integrated Device Electronics" or "Integrated Development Environment," with the former being a hardware name and the latter being a software phrase. Both terms are highly technical, and understanding them will impress even the most nerdy of your friends.

Integrated Device Electronics

The IDE interface is one of the most extensively used hard disk interfaces. The fancy term alludes to how the technology incorporates the drive's electronics controller. While the old IDE standard could only support hard drives with a capacity of up to 540 MB of data, the new EIDE (Enhanced-IDE) standard can accommodate hard drives with capacities of up to 250 GB. It also offers data transfer speeds that are more than twice as fast as the original IDE.

SCSI, which is quicker than EIDE but normally costs more per megabyte, is another common hard drive interface. Newer hard drives may also employ a SATA (Serial ATA) connection, which outperforms both SCSI and IDE in terms of speed and power consumption.

Integrated Development Environment

Computer programming is a time-consuming and difficult undertaking. As a result, software development programs strive to make the software development process as painless as feasible. When creating a software program, an IDE program includes a source code editor, compiler, and, in most cases, a debugger. The IDE manages all project files and provides a centralized interface for authoring source code, connecting files together, and debugging applications.

A runtime environment (RTE) for testing the software may be included in IDE programming software. When a program is run in the RTE, the software can keep track of every event that occurs. This can be a very useful tool while troubleshooting a software. It is simple to make rapid changes to the code, recompile it, and run the program again since the IDE software provides a common interface for authoring and testing the program. Programming is still a difficult task, although IDE software makes the process a little easier.

IEEE

The "Institute of Electrical and Electronics Engineers" is a non-profit organization that creates, defines, and evaluates standards in electronics and computer science. Despite the fact that the IEEE is established in the United States, its standards frequently become international norms. The IEEE 1284 interface (a.k.a. Parallel Port), which is used by many printers, and the IEEE 1394 interface (a.k.a. Firewire), which is a super-fast connection for digital video cameras, hard drives, and other peripherals, are two examples of commonly-used devices standardized by the organization.

IMAP

It's pronounced "eye-map" and stands for "Internet Message Access Protocol." It's a way of getting to e-mail messages on a server without having to download them to your computer. This is the primary distinction between IMAP and another widely used e-mail protocol known as "POP3." Before reading messages, POP3 users must download them to their hard drive. The Benefits of Using IMAP Mail server is that users can check their mail from multiple computers and always see the same messages

Most email clients, such as Microsoft Outlook and Mac OS X Mail, let you choose the protocol your mail server uses. If you utilize your ISP's mail service, inquire as to whether their mail server supports IMAP or POP3 mail. Your e-mail program will not be able to send or receive mail if the protocol configuration is incorrect.

Inbox

The main folder where your incoming mail is stored is called an inbox. Each downloaded message is saved in your inbox, whether you use a webmail interface or an application like Outlook or Mac OS X Mail to check your mail. When you use an e-mail client to check your mail from a POP3 account, the messages are downloaded to the inbox on your local hard drive. If you use an IMAP mail server, however, your inbox is generated on the server, and your messages are also saved there.

Because most people receive more mail than they can handle in a single folder, it's typical to create additional folders to hold your messages. You may either move your messages to other folders you've made (such as "Family," "Friends," "Business," and so on) or trash them by moving them to the Trash once you've read them. It's a good idea to keep the number of messages in your inbox from growing too enormous, regardless of how you manage your mail. Otherwise, you may be forced to declare E-mail Bankruptcy.

Figure 63. Inbox.

Source: Image by Flickr.

Install

Before you can use most software packages, you must first install them on your computer. If you purchase Microsoft Office, for example, you must first install it on your computer before using any of the associated products, such as Word or Excel. Software can be installed using a CD or DVD, an external hard drive, or a networked computer. You can also use a file downloaded from the Internet to install an application or software update.

Software installation is usually a simple procedure. When the license agreement opens up, double-click the installer icon, and then click "I Agree." You may have to select a directory on your hard drive in which to install the software, but the installer will frequently do so for you. You can install certain applications by dragging a folder or application program onto your hard drive. Installing software is a relatively easy process that should not be daunting in any case. You can install your own software if you can make a meal in the microwave.

Integrated Circuit

A little chip called an integrated circuit, or IC, can be used as an amplifier, oscillator, timer, microprocessor, or even computer memory. An integrated circuit (IC) is a compact silicon wafer with hundreds to millions of transistors, resistors, and capacitors. These extremely small devices can use either digital or analog technology to make calculations and store data.

Logic gates, which only work with ones and zeros, are used in digital ICs. A low signal given to a digital IC component will result in a value of 0, while a high signal will result in a value of 1. Digital ICs are the type you'll typically find in computers and networking gear.

Analog, or linear, ICs work with continuous values. This means that a component on a linear IC can accept any value and output any value. Because the output value is a linear function of the input, the term "linear" is employed. A component on a linear IC, for example, might multiply an incoming value by 2.5 and output the result. Audio and radio frequency amplification are common applications for linear ICs.

Intellectual Property

The ownership of intangible and non-physical goods is referred to as intellectual property. It also refers to digital media that may be downloaded via the internet, such as audio and video clips.

Intellectual property is more difficult to defend than other types of property since it is intangible. If tangible property, such as a car, is stolen, it can be recovered or replaced. It may, however, be difficult to retrieve intellectual property if it is stolen. Let's say someone comes up with a brilliant idea for a new invention. If someone else takes the concept, the invention's potential profit may be lost as well.

Similarly, if a digital recording of a new song gets "leaked" on the Internet, thousands of people may download it and resell it. If this occurs, the earning potential of selling music may be significantly reduced.

Intellectual property is frequently used as a legal phrase to protect the rights of creators and innovators because of its monetary consequences. It's also becoming more vital for media production firms who need to protect their digital material's distribution. Innovators and creators can obtain legal protection for their ideas and creations by identifying and creating intellectual property rights.

Interface

"Most digital cameras link to a computer over a common USB interface."

A physical device's interface is a port that allows it to communicate with another device. USB, Firewire, and Ethernet connections are examples of common computer hardware interfaces. Other electronic devices, such as a TV's HDMI ports or a digital piano's MIDI ports, may use various interfaces. A user interface can alternatively be described as a "interface."

Internal Hard Drive

"All of his programs and files are saved on the internal hard drive of the computer," for example.

An internal hard drive is one that is located inside a computer, as the name suggests. Almost every computer has an internal hard drive that serves as the system's primary storage device. The operating system, software applications, and other files are usually found here.

While most computers have only one internal hard drive, some system units and servers feature numerous hard drive slots. Additional internal hard drives can be added to these systems to expand the amount of storage available. If a computer (such as a laptop) only has one internal hard drive bay, additional storage can be added using an external hard drive.

Internet

The Internet, believe it or not, was established by the United States military in 1969, during the Cold War. It was meant to be a "nuke-proof" communication system. Today, the Internet is a global network of thousands of networks and computers that allows millions of people to share information. Long-distance data on the Internet is sent across a network of massive lines known as the Internet backbone. Major Internet service providers such as MCI Worldcom, Sprint, GTE, ANS, and UUNET today sustain the Internet. Because these companies make a lot of money off the Internet, they're compelled to keep their connections stable and quick, which benefits regular Internet users like you and me.

Majority of people believe that the Internet and the World Wide Web are the same thing. They're not! The World Wide Web is what you are browsing right now. It is one of the many features of the Internet. E-mail, FTP, and Instant Messaging are also features of the Internet.

Intranet

This isn't just a typo of "internet," contrary to popular assumption. "Intra" means "internal" or "within," so an Intranet is an internal or private network that can only be accessed within the confines of a company, university, or organization. The difference between the Internet and an Intranet is that "inter" denotes "between or among."

Most organizations employed local networks comprised of pricey proprietary hardware and software for internal communications until recently. Intranets have made internal communication much easier and less expensive by utilizing simple Internet technology. Intranets, like the Internet, use a TCP/IP connection and allow Web browsing. Web sites served within the intranet, on the other hand, can only be accessible by computers linked to the local network. You can go about telling people on the street what you know now that you know the difference between the Internet and an intranet and wow them.

IP (Internet Protocol) Address

This is a code made up of integers separated by three dots that identifies a specific computer on the Internet and is also known as a "IP number" or simply a "IP." To connect to the Internet, any computer, whether it's a Web server or the computer you're using right now, needs an IP address. IP addresses are made up of four groups of numbers ranging from 0 to 255 that are separated by three dots. "66.72.98.236" or "216.239.115.148," for example.

Your Internet Service Provider (ISP) will either assign you a static IP address (which will never change) or a dynamic IP address (which will change) (which changes every time you log on). Dial-up users are often assigned a dynamic IP address each time they log on, reducing the number of IP addresses they must register. If you access the Internet via a network or broadband connection, however, you are more likely to have a static IP address.

IPv4

IPv4 is the fourth edition of the Internet Protocol, and it is now the most widely used version. It uses 32-bit addresses in the format "111.111.111.111." Each sector can have a number between 0 and 255, giving a total of 4,294,967,296 (232) potential addresses. For example: "Most computers display their IP address in the IPv4 format."

Because each machine connected to the Internet requires a unique IP address, 4.3 billion IP addresses are insufficient to meet the global need for unique IP addresses (nice planning guys). As a result, IPv6 is now being developed to replace IPv4, which provides 128-bit IP addresses.

IPv6

An IP address identifies every computer system and device connected to the Internet. IPv4 is the current system for distributing IP addresses. Each machine in this system is given a 32-bit numeric address, such as 120.121.123.124. However, as the number of computers linked to the Internet grows, the number of IP addresses available is expected to run out in a few of years. IPv6 was created to address this issue.

IPv6, often known as IPng (or IP Next Generation), is the IP address system's next planned iteration. (IPv5 was a test version that was primarily utilized for streaming data.) IPv6 employs 128-bit addresses instead of 32-bit addresses, increasing the number of exponential amounts.

For example, IPv4 allows 4,294,967,296 addresses to be used (2^32). IPv6 allows for over 340,000,000,000,000,000,000,000,000,000,000,000,000 IP addresses. That should be enough to last awhile.

Because IPv6 allows for substantially more IP addresses than IPv4, the addresses themselves are more complex. They are typically written in this format:

Hhhh:hhhh:hhhh:hhhh:hhhh:hhhh:hhhh:hhhh

Each "hhhh" segment has a four-digit hexadecimal number, with digits ranging from 0 to 9 and A to F. This is an example of an IPv6 address:

F704:0000:0000:0000:3458:79A2:D08B:4320

The new system also gives extra protection to machines connected to the Internet due to the complexity of IPv6 addresses. Because there are so many IP address possibilities, guessing the IP address of another computer is practically difficult. While IPv6 is supported by the majority of computer systems today, the new Internet protocol has yet to be fully adopted. Computers are frequently allocated both an IPv4 and an IPv6 address during this changeover period. By 2008, the US government has enforced the adoption of IPv6 addresses for all government services, which should hasten the changeover.

Iteration

In a computer program, iteration is the repetition of a function or process. Iterations of functions are common in computer programming because they allow for the processing of several blocks of data in a sequential manner. A "while loop" or "for loop" is commonly used to accomplish this (see the examples below). These loops will keep repeating a process until they reach a given number or case. Iteration is also used in recursive functions, however instead of repeating a process, the entire function is repeated.

While loop: while (x < 30) { … x++; }

For loop: for (x=0; x<30; x++) { … }

A PHP Web page that lists data from a database table is a realistic example of how iterations are employed. The function may write each row in HTML using data from the database until the last row of data is reached in order to show the table on the Web. An iteration of the PHP code would be used to build each row of the table in this example.

Both Web scripts and software programs routinely use iterated functions. They can be simple functions, such as the one shown above, or sophisticated loops with nested loops that may invoke further functions. Even these intricate loops take a fraction of a second to complete, which is incredible.

IVR

It stands for "Interactive Voice Response." IVR stands for interactive voice response, and it is a telephone system that can interpret a mix of touch tone and voice input. It allows people to use their phones to access a database of information. The caller can select from multiple menus of pre-recorded alternatives in a typical IVR system.

While many alternatives are as simple as dialling a phone number, others may need the caller to provide more specific information, such as his name or account number. This information is read by the IVR system and utilized to retrieve the necessary database information.

For example, a bank may have an interactive voice response (IVR) system that allows customers to check their balance or recent transactions by dialling a number. IVR systems are frequently used by credit card issuers and stock brokerage organizations to provide users with account information.

Other applications for the technology include phone surveys, checking movie timings, and contact center forwarding. Using an IVR system is almost like talking to another human being because the caller can audibly respond to pre-recorded messages.

J

Java

Java refers to a programming language produced by Sun Microsystems. Java's syntax is similar to that of C/C++, but it is object-oriented and organized around "classes" rather than functions. Java can also be used to create applets, which are little programs that can be inserted into Web pages. Because it is efficient and simple to use, the language is gaining popularity among Web and software developers.

Java is a "simple, object-oriented, distributed, interpreted, resilient, secure, architecture-neutral, portable, high-performance, multithreaded, dynamic, buzzword-compliant general-purpose programming language," according to Sun Microsystems. It also works like magic when it comes to removing stains.

Figure 64. Java: the most demanded language in IT sector.

Source: Image by Wikimedia.

JavaScript

This is a programming language that, like Java, was created by Sun Microsystems in collaboration with Netscape and can be embedded into regular HTML sites. Even though JavaScript is based on the Java syntax, it is a scripting language that cannot be used to develop standalone programs. Rather, it's mostly utilized to make dynamic, interactive Web pages.

JavaScript can be used to validate form input, produce image rollovers, and open those pesky pop-up windows.

JPEG

The acronym stands for "Joint Photographic Experts Group," which is the name of the group that created the format. But you don't have to remember that because mentioning what JPEG stands for will make even computer specialists think you're crazy. Instead, keep in mind that JPEG is an image file format that is compressed. JPEG images, unlike GIF images, are not confined to a set amount of color.

While JPEG images can include vivid, high-resolution image data, it is a lossy format, meaning that when the image is compressed, some quality is lost. The graphics become clearly "blocky" and some detail is lost if the image is compressed too much. JPEGs, like GIFs, are crossplatform, which means that the same file will look the same on a Mac and a PC.

JRE

It stands for "Java Runtime Environment," and "Java RTE" is another spelling. Java is a programming language in the same vein as C. However, because it is a newer language, not all operating systems support it natively. As a result, Java applets and Java programs may require the installation of a Java Runtime Environment on your computer.

Sun Microsystems (the inventor of Java) developed the JRE, or Java Runtime Environment, which includes the Java Virtual Machine (JVM), code libraries, and components required to run Java programs. The Java Runtime Environment (JRE) is available for a variety of computer platforms, including Mac, Windows, and Unix.

Java programs may not be recognized by the operating system and will not operate if the JRE is not installed. The JRE software creates a runtime environment in which Java programs can be run in the same way that software that has been completely built for the computer's processor can. JRE software is available as a standalone environment and as a Web browser plug-in, allowing Java applications to run in any browser.

JSP

Stands for "Java Server Page." Sun Microsystems created this standard as an alternative to Microsoft's active server page (ASP) technology. JSP pages, like ASP pages, are compiled on the server rather than in the user's browser. After all, they're called "server pages" for a reason. JSP, on the other hand, is based on Java, whilst ASP is based on Visual Basic. JSP pages can be used to create dynamic Web sites and to access database information stored on a Web server. Even though JSP pages may include Java interleaved with HTML, the server parses all Java code. As a result, once the page reaches the browser, it's just HTML. JavaScript, on the other hand, is normally parsed before being used.

Jumper

This is a small metal connector that is used to change hardware configurations and works as an on/off switch. A jumper is made up of two wires and a little metal component. The jumper is turned on after the wires are linked by the metal piece, completing the circuit. The jumper is turned off when the cables are separated. Multiple jumpers, together referred to as a jumper block, are frequently used to tell the computer how to setup a certain device, such as a hard drive or a modem. Motherboards, sound cards, graphics cards, I/O cards, CD-ROM interface boards, modems, and hard drive controller boards, among other things, contain them.

Because manually programming jumpers might be confusing, most gear comes with the appropriate jumpers pre-programmed. If the jumper settings aren't pre-set, there should be documentation with a clear illustration given with the device. Thankfully, a lot of today's plug-and-play technology doesn't require any jumpers. A user-friendly interface on your computer can be used to configure these devices.

K

Kbps

Stands for "Kilobits Per Second." This is not to be confused with kilobytes per second (which is 8 times more data per second). This is a phrase that is frequently used to describe data transfer speeds. 33.6 Kbps and 56 Kbps are two popular modem speeds, for example.

Kernel

Proceed at your own peril because this is a term used by the computing elite. To comprehend what a kernel is, you must first comprehend that today's operating systems are constructed in layers. Serial port access, disk access, memory management, and the user interface are all functions of each layer. The kernel is the operating system's base layer, or the operating system's foundation. The kernel manages memory and offers the most basic "low-level" services, such as hardware-software interface. The operating system will run more efficiently if the kernel is more efficient.

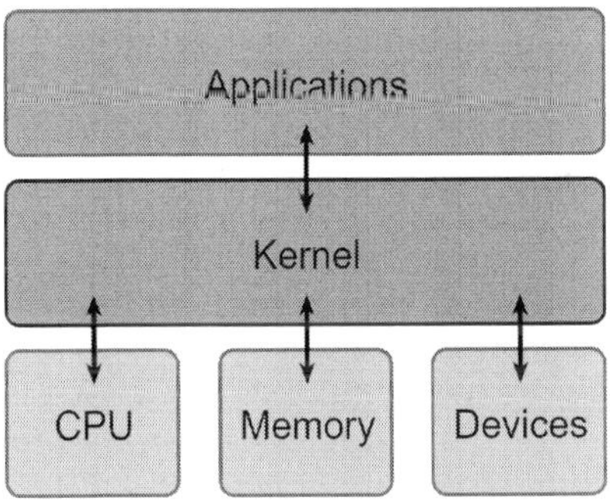

Figure 65. Kernel.

Source: Image by Wikimedia.

Keyboard

A keyboard is, as the name implies, a board of keys. The keyboard, along with the mouse, is one of the most common computer input devices. The design of the keyboard is based on the original typewriter keyboards, which organized letters and numbers in such a way that the type-bars did not jam when typing swiftly. The QWERTY keyboard layout gets its name from the first six letters across in the upper-left-hand corner of the keyboard.

While typewriters inspired the design of computer keyboards, today's keyboards include a variety of different keys. Control, Alt/Option, and Command (Mac) or the Windows key (Windows) can be used as "shortcuts" in conjunction with other keys to accomplish certain functions. For example, pressing Command-S (Mac) or Control-S (Windows) saves a document or project currently open.

A row of function keys (F1 through F16) runs along the top of most modern computer keyboards, together with arrow keys organized in an upside-down T and a numeric keypad on the right-hand side. Some keyboards feature even more buttons, allowing you to adjust the loudness of the system eject a CD, or open programs such as your e-mail or Web browser.

Keyboard Shortcut

A keyboard shortcut is a set of keys that performs a specific action, such as dismissing a window or saving a file. For instance, in a Windows software, pressing "Control-S" or "Command-S" on a Mac is the typical shortcut for saving an open document. On the Mac, you can normally close a window by pressing "Command-W" or in Windows, "Alt-F4." Control-C (Windows) or "Command-C" (Mac) is the shortcut for copying data, and "Control-V" (Windows) or "Command-V" (Mac) is the shortcut for pasting data (Mac).

The majority of keyboard shortcuts are for commands found in a program's menu bar. Most of the commands in the File and Edit menus have standard shortcut. When a command has a keyboard shortcut, the shortcut is normally listed next to the command in the menu. In most Windows programs, the Save option will contain the text "Ctrl+S" next to it, indicating that pressing Control and S simultaneously will save the document.

Keylogger

A keylogger is a computer application that records keystrokes. It accomplishes this by tracking a user's input and preserving a record of all keys hit. The log can be saved to a file or even delivered via a network or the Internet to another system. Keylogger programs are frequently referred to as spyware because they operate without the user's knowledge. Hackers can install them maliciously in order to eavesdrop on what a user is typing. It may be possible to find confidential information such as a login and password combination by looking through the key log data. As a result, if keyloggers are placed on a computer without the user's knowledge, they can pose a serious security concern.

Installing anti-virus or security software that informs you when new programs are installed is the greatest approach to protect yourself against keylogger programs. You must also ensure that no unauthorized individuals gain access to your computer. You may also check the current processes running on your computer on a regular basis to ensure there are no keyloggers or other malware apps operating. Although it is improbable that you have a keylogger program installed on your computer, it is worth checking.

Keywords

Words or phrases that characterize content are referred to as keywords. Images, text documents, database entries, and Web pages can all benefit from metadata. A user can "tag" images or text files with keywords related to their content. These files can later be searched using keywords, making discovering files much easier. A photographer, for example, could use a program like Extensis Portfolio or Apple iPhoto to categorize his nature images with phrases like "nature," "trees," "flowers," "landscape," and so on. He can later find all the photos of flowers by simply searching for the phrase "flowers" after tagging them.

On the Internet, keywords are used in two ways: (1) as search terms for search engines, and (2) as words that identify the website's content.

- Search Engine Search Terms

When you use a search engine to look for something, you enter keywords that tell the search engine what to look for. For example, if you're looking for used vehicles, you may use the keyword "used cars" as your search term. After that, the search engine will return Web pages with content that is related to your search criteria. The more precise (and valuable) your keywords are, the more precise (and useful) the results will be. As a result, if you're looking for a specific used car, try searching for "black Honda Accord used car" to get more precise results.

Many search engines additionally allow Boolean operators, which can be used in conjunction with keywords to narrow down the results. For example, if you just want results about Apple products and not the kind of apples that grow on trees, you might search for "Apple AND computers NOT fruit."

- Web Page Description Terms

The keyword meta tag can also be used to describe the content of a Web page. This tag is found in the HTML head > section of a page and contains words that describe the page's content. The keywords meta tag's aim is to assist search engines in identifying and organizing Web pages, such as the photographs example above. Because webmasters have been known to utilize erroneous tags to boost their search engine rankings, many search engines now assign the keywords meta tag little to no weight when crawling pages.

Kibibyte

A kibibyte is a data storage unit equivalent to 2 to the 10th power, or 1,024 bytes.

Kilobyte

Because it is measured by 210, which equals 1,024, a kilobyte is theoretically 1,024 bytes. Kilobytes, on the other hand, are frequently calculated as 10 to the third power, or 1,000 bytes. While this makes adding kilobytes easy, greater measures can be thrown off by guesswork. This is because one megabyte is equal to 1,024 kilobytes, and one gigabyte is equal to 1,024 megabytes, and so on.

Kilobytes are the unit of measurement for most little files on your computer. Thumbnail images, for example, may only take up 5 to 10KB of space. A larger JPEG image with 900x600 pixels can take up 250KB of space. Text files are frequently under 1KB in size. The majority of the documents you store on your computer should be between 1 and 1,024 KB in size. Megabytes are used to measure anything larger than 1,024KB.

L

LAN

A local area network, or LAN, is a computer network that is limited to a small area, such as an office building, university, or even a private residence. LANs are used in most mid- to large-sized enterprises nowadays, making it easy for employees to share information. Ethernet-based LANs, which employ Novell or Oracle software, are currently the most common type of LAN. Wireless LANs, on the other hand, have become a popular alternative with the introduction of wireless networking.

Latency

This is the time it takes for a data packet to travel over a network connection. There is "latent" time when a packet is sent, while the computer that sent the packet waits for confirmation that it has been received. The two parameters that govern the speed of your network connection are latency and bandwidth.

LCD

"Liquid Crystal Display" is the abbreviation for "liquid crystal display." LCDs are ultra-thin displays seen in laptop computers and flat panel monitors. Handheld TVs, PDAs, and portable video game devices all use smaller LCDs. An electrically reactive material is sandwiched between two electrodes to form the image on an LCD panel. Increasing or decreasing the electrical current can modify the color of this material. Due to the fact that LCD panels are based on the principle of blocking light (rather than emitting it), they use up much less power than standard CRT (Cathode-Ray Tube) monitors.

LDAP

Stands for "Lightweight Directory Access Protocol." This is the method to use if you wish to make directory information available via the Internet. LDAP is a condensed version of the previous X.500 directory standard. The fact that LDAP operates well over TCP/IP networks (unlike X.500) means that information can be accessible by anybody with an Internet connection. It's also an open protocol, meaning that directories can be stored on any machine (i.e., Windows 2000, Red Hat Linux, Mac OS X).

Here are the several layers of a simple LDAP tree hierarchy to give you an understanding of how an LDAP directory is organized:

- The top-level directory
- Countries
- Organizations
- Divisions, departments, and so forth
- Individuals
- Personal resources like files and printers.

The majority of LDAP communication happens behind the scenes; thus, the average user won't see it when surfing the web. It is, nonetheless, an important technology to be aware of. At the very least, it's a new term to impress your folks with.

Leaf

What do a computer's hard disk and a towering oak have in common? While the hard drive may not appear to be organized in the same way as a tree from the outside, it is. On a hard drive, files and folders are divided into branches, with each directory acting as a branch with files and folders. The branches are made up of folders, and the leaves are made up of files. As a result, a leaf is a file that is contained within a directory on your hard disk.

A leaf is a node in a tree that has no child nodes in technical terms. Because files, unlike folders, cannot contain child nodes, they are always leafs. A leaf can also be referred to as a leaf node when referring to a tree structure.

LED

"Many modern cars have LED brake lights," for example.

The term "Light-Emitting Diode" refers to an electronic light source that is turned on by an electrical current. LEDs save a lot of energy and last a lot longer than traditional light bulbs. Indicator lights, clock displays, street lights, and a variety of other applications can all benefit from them. In LCD displays and televisions, LEDs have begun to replace conventional bulbs as the backlight source.

LIFO

LIFO stands for "Last In, First Out," a data processing method in which the most recent items input is the first to be discarded. FIFO (First In, First Out) is the polar opposite of LIFO, in which objects are deleted in the order they were entered. Imagine stacking a deck of cards by laying one card on top of the other, starting at the bottom, to better comprehend LIFO. You begin removing cards from the top of the deck after it has been entirely stacked. Because the last cards to be placed on the deck are the first to be removed, this procedure is an example of the LIFO approach.

When pulling data from an array or data buffer, computers sometimes employ the LIFO approach. The LIFO method is used when a computer needs to access the most recent data entered. The FIFO approach is utilized when data must be retrieved in the order it was entered.

Link

When you're surfing the web and come across a highlighted word or phrase, it's likely that you're looking at a link. You can "jump" to a new web page or an entirely different Web site by clicking on a link. Text links are usually blue and underlined, but they can be any color and do not have to be. Images can also be used to direct visitors to other websites. When you hover your mouse over a link on a Web page, the arrow transforms into a small hand, indicating that it is a link. The term "hypertext" refers to how links can swiftly redirect you to another Web page.

Linux

"The Linux operating system is extensively used on web servers."

Linux is a Unix-like operating system that is free and open source. CentOS, Debian, Ubuntu, and Red Hat are just a few of the distributions that offer it. Linux can run on a variety of hardware platforms and can even be used as the operating system for devices other than computers. Despite the fact that Linux is not frequently installed on personal computers, it is widely utilized by Web hosting firms and has a wide range of other specialized uses.

LISTSERV

This phrase appears to be missing a "e," although it is spelled correctly. A listserv, often known as a list server, is a simple application that delivers messages to several e-mail addresses automatically. When someone joins a mailing list, the listserv adds their address to the list and sends future e-mail messages to that address as well as all the others on the list. When a person unsubscribes, the address is simply removed from the listserv. That is, at least, how it is meant to function. Unfortunately, unsubscribing from some SPAM lists merely adds you to new lists.

Load Balancing

It is a term used to describe the process of balancing. Computer networks are sophisticated systems that handle hundreds, thousands, or even millions of data packets per second. As a result, it is critical that data be routed efficiently in order for networks to handle massive amounts of data. For instance, if a network has ten routers and two of them are is not working very efficiently because they are doing 95% of the work. If each router handled around 10% of the traffic, the network would run substantially quicker.

Similarly, if a website receives hundreds of views per second, splitting the traffic over numerous Web servers is more efficient than relying on a single server to manage the entire load. Load balancing improves network efficiency. It equally distributes processing and traffic throughout a network to ensure that no single device is overburdened. Load balancing is a technique used by web servers to distribute traffic load evenly among multiple servers, as shown in the example above. This allows them to use the available bandwidth more effectively, and therefore provides faster access to the websites they host.

Load balancing, whether on a local network or on a large Web server, necessitates the use of hardware or software to distribute incoming traffic among the available servers. One or more servers may be dedicated to balancing the load among the other servers and devices in a network that receives a lot of traffic. Load balancers are the name given to these servers, which is not surprising.

Load balancing is also used by clusters, or multiple computers that work together, to distribute processing duties across the available systems.

Localhost

The term "localhost" refers to the machine on which a software is running. If you're using a Web browser on your computer, for example, your computer is considered the "localhost." When utilizing a single computer, the localhost does not need to be supplied; however, when using multiple computers, the localhost must be specified. A network administrator, for example, might start a Web server on one system while using a remote access tool on another. These applications would be executed on machines other than the localhost.

The two non-local machines in the preceding example must be identified by their IP addresses. The local machine's IP address is 127.0.0.1, which is defined as "localhost." The information supplied to this address is sent back to the local machine, making it a "loopback" address. When indicating what server, the code should run on or where a database is hosted, localhost is frequently used in Web scripting languages like PHP and ASP.

Log file

"The list of installed files is kept in a log file," for example. A log file is a file that keeps track of events that are generated by software or output by hardware. The majority of log files are saved in plain text format, which may be read using a simple text editor.

Logic Gate

Logic gates are the essential building blocks of digital integrated circuits, performing basic logical tasks. The majority of logic gates take two binary values as input and output a single value of 1 or 0. Some circuits have only a few logic gates, while others, like microprocessors, have millions. There are seven different types of logic gates, as shown in the diagram below.

Except for the NOT gate, each logic gate in the following examples has two inputs, A and B, which can be either 1 (True) or 0 (False) (False). If the result is positive, the output is a single value of 1; if the result is false, the output is a single value of 0. In the following examples, each logic gate except the NOT gate has two inputs, A and B, which can either be 1 (True) or 0 (False). The resulting output is a single value of 1 if the result is true, or 0 if the result is false.

AND – True if A and B are both True

OR – True if either A or B are True

NOT – Inverts value: True if input is False; False if input is True

XOR – True if either A or B are True, but False if both are True

NAND – AND followed by NOT: False only if A and B are both True

NOR – OR followed by NOT: True only if A and B are both False

XNOR – XOR followed by NOT: True if A and B are both True or both False

It is feasible to conduct extremely complicated operations by merging thousands or millions of logic gates. The size of the chip divided by the size of the logic gates determines the maximum number of logic gates on an integrated circuit. Smaller transistors mean more sophisticated and quicker processors since transistors make up the majority of the logic gates in computer processors.

Login

You are being asked to submit your login information if you are ever requested to enter your username and password. A login is a collection of information that confirms your identity. This could be a username and password or a security code and ID number. Many secure websites employ login information to verify visitors' identities before granting them access to restricted areas. This term, unlike "brush" and "comb," should not be used as a noun and a verb at the same time. It should only be used as a noun (you log in to a server, not into it).

Lossless

Lossless compression reduces the size of a file without sacrificing quality. Both image and audio files can benefit from this seemingly magical approach of file compression. Newer compression methods, such as JPEG 2000 and Apple Lossless compression, can be used to create lossless compressed files, whereas JPEGs and MP3s employ lossy compression.

Lossless compression essentially rewrites the original file's data in a more efficient manner. However, because no quality is sacrificed, the resulting files are usually significantly bigger than lossy compressed image and audio files. Lossy compression, for example, can reduce a file to a tenth of its original size, whereas lossless compression is unlikely to reduce a file to half its original size.

Lossy

Data and quality from the original version are lost when files are compressed in a lossy manner. Lossy compression is most commonly linked with picture files like JPEGs, but it can also be employed with audio files like MP3s or AACs. The "lossiness" of an image file might be a problem.

Lossyness in audio files can cause a watery sound or limit the dynamic range of the music.

Because lossy compression removes data from the original file, the resultant file often takes up significantly less disk space. A JPEG image, for example, can reduce the file size of an image by more than 80% with little perceptible effect. A compressed MP3 file, for example, may be one tenth the size of the original audio file while sounding nearly identical.

"Almost" is the essential word here. Both JPEG and MP3 compression eliminate data from the original file, which may be seen if you look closely. Both of these compression methods have a number of "quality parameters" that control how well the data is compressed. The quality option is a trade-off between file size and quality. Although a file with higher compression takes up less space, it may not look or sound as good as a file with lower compression. Some image and music formats support lossless compression, which preserves the quality of the file.

LPI

It stands for "Lines Per Inch." LPI is a metric for determining the resolution of halftone images. Due to the fact that halftone images are produced as a series of dots, the higher the LPI value, the more dense the dots can be, resulting in finer resolution. Newspapers have an average resolution of 85 lpi, but magazines might have a resolution of 133 lpi or greater. You're more likely to see halftone dots in newspaper print than in magazines because the naked eye can identify them up to 120 lpi. of course, if you look closely enough, the dots in photographs printed at 150 lpi or above may be visible. However, even at 85 lpi, it is reasonable to regard the dots as a continuous image in normal viewing.

M

MAC

Address stands for "Media Access Control Address" and it has nothing to do with Apple computers. A MAC address is a hardware identifying number that allows each device on a network to be uniquely identified. Every network card, such as an Ethernet or Wi-Fi card, comes with a unique MAC address that cannot be modified.

Because there are millions of networkable devices, each of which requires its own MAC address, there must be a large number of potential addresses. MAC addresses are therefore composed of six two-digit hexadecimal digits separated by colons. A MAC address of 00:0d:83:b1:c0:8e, for example, might be assigned to an Ethernet device. Fortunately, you won't need to know this address because most networks will detect it automatically.

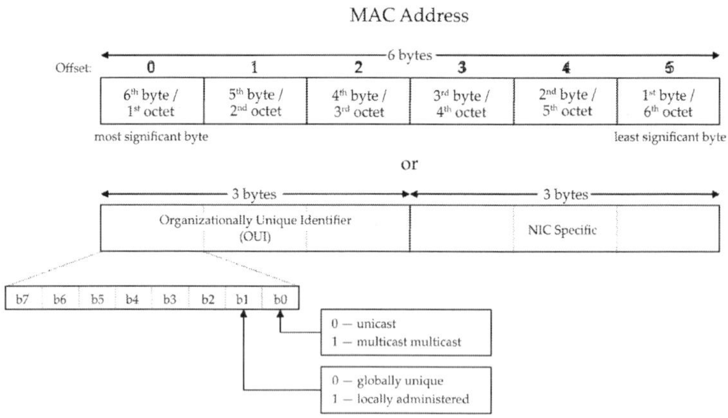

Figure 66. Media access control address.

Source: Image by Wikimedia Commons.

Macro

Macro has two computer-related definitions, both of which entail improving the efficiency of your computer experience.

- A Shortcut on the Keyboard

The terms "macro" and "keyboard shortcut" are frequently interchanged. Keyboard shortcuts are key combinations that accomplish tasks like saving a file, dismissing a window, or copying and pasting information. View the pages below for a list of keyboard shortcuts for Mac and Windows:

1. Keyboard Shortcuts for Mac OS X; and
2. Keyboard Shortcuts for Windows XP.

- A Small Program

A macro is a short program (or script) that automates repetitive processes. Scripts are typically run within programs and can be written by the user. For instance, a user may program a macro in Microsoft Word to insert his complete address when he presses a specific key combination. A Microsoft Excel user could create a macro to format data in a spreadsheet's selected column.

Although Word and Excel make it simple to build custom macros, numerous other programs also allow users to do so. Not all applications, however, refer to automated commands as macros can save you a lot of time by automating repetitive tasks. If you find yourself doing the same thing over and over again when using a certain program, you may want to simplify the process by recording a macro.

Mainframe

A mainframe computer is a supercomputer designed for high-volume, processor-intensive computation. Large corporations and research institutions are the most common users. A mainframe is unlikely to be found in any home. Mainframes are just behind supercomputers, the world's most powerful computers, in the computer hierarchy. (They're called "supercomputers" for a reason.) A mainframe, on the other hand, can normally run multiple applications at once at fast speeds, whereas supercomputers are meant for a single task. Currently, IBM and Unisys are the two largest mainframe manufacturers.

Malware

Malware, short for "malicious software," refers to computer programs that are designed to cause harm or do other undesirable behaviors. "Mal" is a prefix in Spanish that meaning "bad," therefore the word "badware" is an excellent way to remember it (even if you don't speak the language).

Viruses, worms, trojan horses, and spyware are all types of malware. Viruses, for example, can destroy files and directory information on a computer's hard disk. Spyware can collect information from a user's computer without the user's knowledge. This can range from the Web pages that a user accesses to sensitive information like credit card numbers.

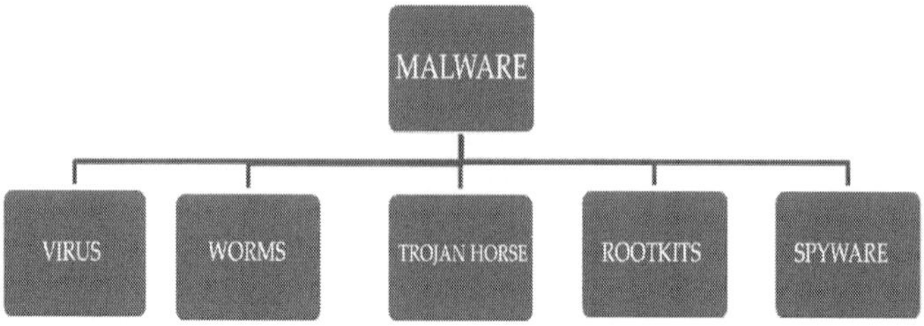

Figure 67. Types of malware.

Source: Image by Wikimedia.

It's sad that there are malicious software programmers out there, but it's good to be aware of the possibility. You may use anti-virus and anti-spyware software to discover and remove harmful apps from your computer. So join the fight against malware by installing some anti-malware software on your computer!

MANET

Stands for "Mobile Ad Hoc Network." A MANET is an ad hoc network that can move around and reconfigure itself as needed. Because MANETs are movable, they link to numerous networks via wireless connections. This could be a normal Wi-Fi connection or another type of transmission, such as cellular or satellite.

Some MANETs are restricted to a local area of wireless devices (such as a group of laptop computers), while others may be connected to the Internet. A VANET (Vehicular Ad Hoc Network), for example, is a sort of MANET that allows cars to communicate with roadside infrastructure. While the vehicles may not have a direct Internet connection, the wireless roadside equipment may be, allowing data from the vehicles to be transferred over the Internet. The vehicle data might be used to track freight fleets or analyze traffic conditions. Because MANETs are inherently insecure due to their dynamic nature, it is critical to be vigilant about what data is transferred across them.

Markup Language

A markup language is a syntax for specifying elements within a document, such as a web page or a data file. Tags are used to designate the start and end of each element. The code in markup files can be read and updated in a text editor because they are saved as plain text files. For example: "HTML is a markup language used to define the content of webpages."

HTML, SGML, and XML are popular markup languages. The layout of webpages, as well as the elements within each page, are defined in HTML, which is a subset of SGML (the Standard Generalized Markup Language). A section on an HTML page often has a title, metadata, and references to one or more CSS files. The content of an HTML page is defined by the section, which might include text, picture references, and links to other pages.

Custom tags are used in XML files to define elements and sub-elements inside a document. XML files are used to store data in a structured fashion, whereas HTML files are used to format how information is shown. In an XML data file, for example, a tag may be used to define an employee element. To define the employee's name and position, the tags name > and title > could be used as sub-elements. XML is a popular choice for transferring structured data across programs because of its ease and versatility.

Mashup

The term "mashup" can refer to a variety of things. It was first used to describe songs that combined two different musical styles into a single song. A mashup, for example, could be a classic rock song set to a well-known hip-hop beat. It's also used to describe videos that were put together from a variety of clips from various sources. A video mashup, for example, is a skating film made out of numerous different skateboard films discovered online.

A mashup is also a term for a Web application that integrates numerous services into one. For example, a Web forum might have a mashup that uses Google Maps to show where users are posting from around the world. Yahoo! Pipes, a mashup that aggregates RSS feeds into a single page that can be accessed using a graphical interface, is available.

The basic goal of most Web mashups is to combine data into a user-friendly interface. The possibilities of mashups are as endless as the combinations of Web applications. As a result, don't be shocked if you find mashups on your favorite websites as they continue to expand.

Maximize

When you maximize a window, it expands. On a Mac, a maximized window spans the entire screen, whereas on Windows, it just takes up as much space as is required. In Windows, the maximize button is adjacent to the close button in the upper-right corner of the window. When you double-click the title bar, the window will be maximized as well. On a Mac, the maximize button is three green buttons located next to the red and yellow buttons in the upper-left corner of the window. For example: "If you need to view more of the window's contents at one time, try maximizing the window."

To hide an open window, you can click the minimize button, which is located next to the maximize button on both Macintosh and Windows computers.

Mbps

Stands for "Megabits Per Second." One megabit (1,000 kilobits) is equivalent to one million bits. While the terms "megabit" and "megabyte" sound similar, a megabit is roughly one eighth the size of a megabyte (since there are eight bits in a byte). Data transmission speeds of high bandwidth connections, such as Ethernet and cable modems, are measured in megabits per second (Mbps).

MBR

An MBR, or "master boot record," is a particular region of a hard disc that holds the boot code and defines the disc partitions. It is only 512 bytes in size and is found in the first sector of the disc. For example: "Microsoft's Fdisk utility can be used to create and update the master boot record."

Master boot records (MBRs) were first launched in the early 1980s and have been utilized by DOS-compatible systems for more than two decades. MBRs, on the other hand, can only designate up to 2TB of space, therefore they can't be used with storage devices greater than that. As a result, most computers today use the GPT (GUID partition table), which has no limit of two terabytes.

MCA

Stands for "Micro Channel Architecture." It's an IBM-designed expansion bus that was utilized in the company's PS/2 desktop computers. An expansion bus connects additional cards to the computer's motherboard, increasing the number of I/O ports available. SCSI, USB, Firewire, AGP, and DVI connectors, among others, are available.

The MCA standard was created to replace the AT and ISA buses that had previously been used in IBM PC/AT compatible computers. Despite the fact that the MCA bus architecture was smaller and faster than AT and ISA, IBM kept it as a proprietary standard. Other manufacturers were hesitant to adopt the standard because the MCA architecture was incompatible with other standards. Most PC companies incorporated the more universal PCI and AGP expansion buses into their computers, which are both widely used today.

Mebibyte

A mebibyte is a unit of data storage that equals 2 to the 20th power, or 1,048,576 bytes. While a megabyte can be estimated as 10^6 or 1,000,000 bytes, a mebibyte is exactly 1,048,576 bytes. This is to avoid the ambiguity associated with the size of megabytes. A mebibyte is 1,024 kibibytes and precedes the gibibyte unit of measurement.

Media

In general, the term "media" refers to a variety of communication methods. Television, radio, and newspapers, for example, are all diverse sorts of media. The phrase can also be used to refer to the press or news organizations as a whole. "Media" is also used as a collective term in the computer sector; however, it refers to several forms of data storage possibilities.

Hard drives, portable drives (such as Zip discs), CD-ROM or CD-R discs, DVDs, flash memory, USB drives, and yes, floppy discs are all examples of computer media. If you wish to take pictures from your digital camera to a photo processing store, for example, they may ask you what kind of media your photos are saved on. Are they on the flash memory card inside your camera or are they on a CD or USB drive? For this and many other reasons, it is helpful to have a basic understanding of what the different types of media are.

Megabyte

A megabyte is 2 to the 20th power, or 1,048,576 bytes. 10 to the 6th power, or one million (1,000,000) bytes, is an estimate. A megabyte is 1,024 kilobytes in size and comes before the gigabyte. Megabytes are the standard unit of measurement for large computer files. A high-quality JPEG photo taken with a 6.3 megapixel digital camera, for example, takes up around 3MB of space. CDs can carry up to 700MB of data, thus a four-minute CD-quality audio clip takes up around 40MB of space.

Megahertz

One million cycles per second is one megahertz. It's used to determine how fast electronic gadgets transmit data. This is the most common spot you'll come across. Megahertz is a unit of measurement for CPU clock speed. For example, an 800 Mhz Pentium III processor.

It's vital to understand that megahertz simply refers to the processor's clock speed (the number of cycles it can manage per second). not the ultimate achievement Because megahertz only measures a particular component of the CPU, it's possible that a processor with a slightly higher megahertz reading is faster. A Mac with a 500 MHz PowerPC G4 processor, for example, can execute some calculations faster than a PC with an 800 MHz Pentium III processor. Because the G4 can process more instructions per clock cycle than the Pentium, it outperforms the Pentium. This is a point that Mac users frequently emphasize, however it is no longer relevant because Macintosh machines now employ Intel processors as well.

Megapixel

One million pixels make up to a megapixel. It's a term that's widely used to define how high a digital camera's resolution is. A 7.2 megapixel camera, for example, can capture around 7,200,000 pixels. The more megapixels a camera has, the more detail it can record. As a result, while purchasing a digital camera, the megapixel count is an important feature to consider.

The number of vertical pixels captured by the camera's sensor, or CCD, is multiplied by the number of horizontal pixels captured by the camera's sensor, or CCD. The original Canon Digital Rebel, for example, had a total of 6,291,456 pixels, with 2048 vertical by 3072 horizontal pixels (2048 x 3072). As a result, it's thought to feature a 6.3 megapixel camera. The Sony T10 has a resolution of 3072 x 2304 pixels, for a total of 7,077,888 pixels, making it a 7.2 megapixel camera (because not all the pixels are used).

Megapixels are useful in selling digital cameras since "6.3 megapixels" is easier to say than "6,291,456 pixels." It's also a little simpler to recall. While megapixels are crucial, it is also beneficial to understand a camera's other specs. Shutter speed, shooting modes, start-up time, flash quality, and color correctness, for example, can all have a significant impact on the camera's performance.

After all, it makes little difference how many megapixels your camera has if all of your photos are blurry and have terrible color. As a result, while you should examine the megapixel count on a camera before purchasing it, you should also look at the other specifications.

Memory

Computers, like humans, rely heavily on memory. They must handle and store data in the same way that people do. On the other hand, computers save data in digital format, which implies that the information can always be retrieved in the same format as it was saved. The computer's memory, unlike ours, does not deteriorate with time.

While memory can refer to any type of data storage device, it is most commonly associated with RAM, or random-access memory. When your computer turns on, it loads the operating system into RAM, or random-access memory. Because the event handlers are all loaded into RAM, your computer can access system operations like managing mouse clicks and keystrokes. Whenever you open a program, the interface and functions used by that program are also loaded into RAM.

Memory Module

A RAM chip is also known as a memory module. It is frequently used as a catch-all word for SIMM, DIMM, and SO-DIMM memory. While memory modules come in a variety of shapes and sizes, they all serve the same purpose: to store temporary data while the computer is functioning.

Memory modules are available in a variety of sizes and pin layouts. The first SIMMs, for example, had 30 pins (which are metal contacts that connect to the motherboard). However, modern SIMM chips contain 72 pins. DIMMs generally come in 168-pin layouts, while some DIMMs contain as many as 240 pins. SO-DIMMs have a smaller form factor than standard DIMM chips, and come in 72-pin, 144-pin, and 200-pin configurations.

While the official name for computer memory is "memory module," the terms "RAM," "memory," and "RAM chip" are also acceptable. But keep in mind that while memory words are interchangeable, memory is not. Because most computers only accept one type of memory, this is the case. As a result, if you decide to update your computer's RAM, ensure the memory modules you purchase are compatible with your machine.

Memory Stick

Sony created the Memory Stick flash memory format. It stores data for digital cameras, camcorders, and other electronic devices. Because Memory Stick is a Sony-only device, it is used by practically all of the company's flash-based products. Unfortunately, this means Memory Stick cards are incompatible with the vast majority of non-Sony products.

Memory Stick cards are available in two versions: Memory Stick PRO and Memory Stick PRO Duo. Memory Stick PRO cards are 50 mm long by 21.5 mm wide and are 2.8 mm thick. Memory Stick PRO Duo cards are 31 mm long by 20 mm wide and are only 1.6 mm thick. High-speed versions of Memory Stick media support data transfer rates up to 80 Mbps, or 10 MB/sec, which is fast enough to record high-quality digital video.

Meta Search Engine

Scan engines that search other search engines are known as meta search engines. Confused? Simply put, a meta search engine sends your query to a number of different search engines and then offers a summary of the results. As a result, the search results you see are a compilation of many queries.

While this method broadens the breadth of your search compared to using a single search engine, the results are not necessarily superior. This is because the meta search engine must choose the best results from several search engines using its own algorithm. A meta search engine's results are frequently less relevant than those of a traditional search engine.

Metadata

Other data is described through metadata. It gives details about the contents of a specific object. An image may contain metadata that defines the size of the image, the color depth, the image resolution, the date the image was made, and other information. The metadata of a text document may include information such as the document's length, author, date of creation, and a brief summary. Metatags are commonly used to include info on web pages. Meta tags for description and keywords are often used to describe the content of a Web page. When adding pages to their search index, most search engines use this information.

MIDI

Stands for "Musical Instrument Digital Interface." It's a connection standard that allows musicians to link musical instruments (such keyboards and synthesizers) to computers. A musician can easily produce and alter digital music tracks using MIDI. The notes played, the length of the notes, the dynamics (volume changes), the tempo, the instrument being played, and hundreds of other factors, known as control changes, are all recorded by the MIDI system.

Because each note is digitally recorded in MIDI, editing a track of MIDI music is significantly easier and more precise than editing an audio file. With the push of a button, the musician can alter the notes, dynamics, tempo, and even the instrument being performed. Furthermore, because MIDI files are essentially text documents, they occupy extremely little disc space. The only catch is that to record and playback MIDI files, you'll need MIDI-compatible devices or software.

Mnemonic

A mnemonic (pronounced "nemonic") is a pattern that can be used to help people remember things. This pattern is usually made up of letters or words. To help music students recall the notes of the staff, such as E, G, B, D, and F, the slogan "Every Good Boy Does Fine" might be utilized. The name "Roy G. Biv" is often used to memorize the order of colors in a rainbow (or other light spectrum) – Red, Orange, Yellow, Green, Blue, Indigo, Violet.

While word initials are frequently used as mnemonic devices, rhyming words and poems can also be utilized to remember information. Furthermore, visuals can be linked to words or phrases to aid with their recall. Mnemonics help people categorize information better, which makes it simpler to recall, because the human brain organizes information in "chunks."

Modifier Key

When two keys are pressed together, a modifier key changes the operation of the other key. Shift, Control, Alt, Command, Option, and Function are all common modifier keys. On Windows keyboards, the Control (Ctrl) and Alt keys are commonly seen, whereas on Mac keyboards, the Command (Cmd) and Option keys are commonly found.

All computers have a Shift key that is used to capitalize letters and enter symbols (such as the ones found above the number keys). Keyboard shortcuts are commonly entered using the Control and Command keys. For example: "Modifier keys are commonly used to enter keyboard shortcuts."

Monitor

The terms "monitor" and "computer screen" or "display" are frequently interchanged. The monitor shows the user interface and open programs on the computer, allowing the user to interact with it using the keyboard and mouse.

Cathode ray tubes (CRTs) were utilized in older computer displays, which made them quite hefty and took up a lot of desk space. The majority of current monitors use LCD technology and are referred to as flat screen displays. These slim monitors take up a fraction of the space of previous CRT displays. This means people with LCD monitors have more desk space to clutter up with stacks of papers, pens, and other objects.

Figure 68. Monitor or a computer screen.

Source: Image by pixnio.com.

It's also possible to use "monitor" as a verb. A network administrator can monitor network traffic, which means he looks at it to see whether bandwidth utilization is within a specific range and to see if any external sources are seeking to connect to the network. Software packages can keep track of the system's CPU, RAM, and hard drive use.

Finally, speakers utilized for sound monitoring are referred to as monitors. To listen to recordings, audio engineers generally use "studio monitors." Engineers can correctly mix and master audio records with these high-end speakers. So, a sound mixer may be visually viewing a recording on a computer monitor while also monitoring the sound on audio monitors. As you are aware, "monitor" serves as a rather multipurpose word.

Moodle

Stands for "Modular Object-Oriented Dynamic Learning Environment." Martin Dougiamas created Moodle, which is an open-source course administration system. Thousands of educational institutions all over the world utilize it to provide an organized interface for e-learning, or online learning.

Educators can use Moodle to develop online courses that students can access as a virtual classroom. A typical Moodle home page will have a list of participants (including the teacher and students), as well as a calendar with the course schedule and assignment list. Online quizzes, forums where students can post comments and ask questions, glossaries of terminology, and links are some of the other Moodle features.

Motherboard

The motherboard, often known as the mainboard or logic board, is the main circuit board of your computer. The motherboard is the largest piece of silicon visible when you open your computer. The CPU, ROM, memory RAM expansion slots, PCI slots, and USB ports are all attached to the motherboard. Controllers for devices such as the hard disc, DVD drive, keyboard, and mouse are also included. Basically, the motherboard is what makes everything in your computer work together.

The chipset is a collection of chips and controllers found on each motherboard. New chipsets are frequently used in the development of new motherboards. The good news is that these boards are usually more efficient and quicker than their forerunners. The bad news is that new chipsets frequently do not operate with older components. of course, it may be more cost-effective to just buy a new computer if you are planning on upgrading multiple components.

MySQL

MySQL is an open-source relational database management system that is pronounced "My S-Q-L" or "My Sequel." It is built on the structured query language (SQL), which is used to add, delete, and alter data in databases.

MySQL supports standard SQL commands like ADD, DROP, INSERT, and UPDATE. MySQL is a database management system that can be used for a variety of purposes, but it is most typically seen on Web servers. Web pages that access information from a database may be found on a MySQL-powered website. The content of these websites is frequently referred to as "dynamic," which means that the material is created from a database as the page loads. Database-driven websites are those that make extensive use of dynamic web pages.

Many MySQL-powered websites additionally make use of a Web scripting language like PHP to access data from the database. MySQL commands can be implemented into PHP code, enabling the generation of part or all of a Web page from database data. Since both MySQL and PHP are both open source (meaning they are free to download and use), the PHP/MySQL combination has become a popular choice for database-driven websites.

N

Name Server

A name server is a computer that converts domain names into IP addresses. This allows users to access a website by entering in the domain name rather than the website's real IP address. When you input "www.microsoft.com," for example, the request is forwarded to Microsoft's name server, which provides the Microsoft website's IP address.

Each domain name must have at least two name servers listed when the domain is registered. These name servers are usually named ns1. servername.com and ns2.servername.com, where "server name" is the name of the server. The first server listed is the primary server, on the other hand the second is used as a backup server if the first server is not responding.

Domain Name System (DNS) relies heavily on name servers. They enable websites to utilize domain names rather than IP addresses, which are significantly more difficult to remember. A WHOIS lookup tool can be used to determine the name servers for a specific domain name.

Netmask

A netmask is a set of IP addresses that is used to specify a range of addresses. It is similar to a subnet mask but it is used to define classes of IPs rather than a range of IPs that may be used within a network. Class B IP addresses, for instance, have a subnet mask of 255.255.0.0. This means that the IPs must all have the same first two sections, but the second two sections may have different numbers. For example: "Class A IP addresses have a netmask of 255.0.0.0."

Due to a scarcity of IP addresses, most IP ranges are assigned to Class C, which has the netmask 255.255.255.0. This specifies a set of IP addresses that have the same values in the first three parts but can have any number between 0 and 255 in the fourth. As a result, a Class C netmask specifies a range of 256 addresses.

Network

A network is created when two or more computers are connected to each other. The goal of a network is to make it possible for various systems to share files and information. Internet is a worldwide network of networks. It is possible to connect computer networks through cables (Ethernet cables or phone lines) or through wireless networking cards that send and receive data in a wireless manner.

Network Topology

The arrangement of systems on a computer network is referred to as network topology. It specifies how the network's computers, or nodes, are grouped and connected to one another. Star, ring, line, bus, and tree layouts are some of the most prevalent network topologies. The following are the topologies:

- Star – One central node is connected to each of the other nodes on a network. Similar to a hub connected to the spokes in a wheel.
- Ring – Each node is connected to exactly two other nodes, forming a ring. Can be visualized as a circular configuration. Requires at least three nodes.
- Line – Nodes are arranged in a line, where most nodes are connected to two other nodes. However, the first and last nodes are not connected as if they were in a ring.
- Bus – Each node is connected to a central bus that runs along the entire network. All information transmitted across the bus can be received by any system in the network.
- Tree – One "root" node connects to other nodes, which in turn connect to other nodes, forming a tree structure. Information from the root node may have to pass through other nodes to reach the end nodes.
- When putting together a network, a network administrator should be aware of the advantages and disadvantages of various network topologies. The administrator can choose the configuration that is most efficient for the network's intended purpose by weighing the benefits of each type.

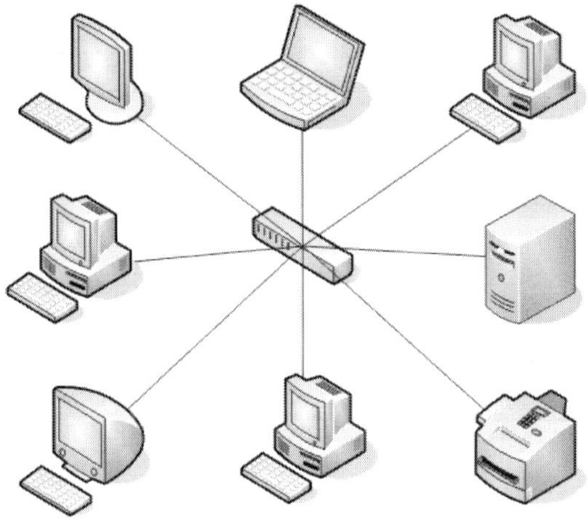

Figure 69. Star topology: A type of network topology.

Source: Image by Wikimedia.

Node

A node is any system or device that is connected to a network. There are eight nodes on a network if it connects a file server, five PCs, and two printers, for example. Every network device has a network address, such as a MAC address, that uniquely identifies it. This makes it easier to keep track of where data is being transported on the network.

A leaf, which is a folder or file on your hard disc, is sometimes referred to as a node. A node, also known as a nodal point, is a point in physics where numerous waves converge to create a net amplitude of zero.

NTFS

NTFS is the abbreviation for "New Technology File System." Microsoft debuted the NTFS file system with Windows NT, and it is supported by later versions of Windows, such as Windows 2000 and Windows XP. Over the previous file system, FAT32, NTFS has a lot of improvements (File Allocation Table). One of the most significant advantages of NTFS is that it has features that enhance reliability. Fault tolerance, for example, is included into the new technology file system, which automatically repairs hard drive defects without presenting error warnings. It also maintains thorough transaction logs for tracking hard disc problems. This can assist prevent hard disc failures while also allowing you to recover files if your hard drive does fail.

Permissions (such as read, write, and execute) can also be adjusted for specific directories and files using NTFS. It even provides for spanning volumes, which enables for the distribution of file directories across many hard drives. The only reason you wouldn't want to format your hard disc using NTFS is if you want slow, obsolete technology or if you need to run an older operating system like Windows 95 or MS-DOS. of course, if you're still using DOS, there's a decent chance you enjoy archaic technology.

Null

A variable is deemed null when it has no value. Because 0 is a real value, having a null value is not the same as having a value of 0. In a Boolean test, however, both null and zero provide a FALSE result. Boolean tests are commonly used by programmers to determine whether a variable has been assigned a value.

Null Character

A null character is one in which all bits are set to zero. As a result, it has a numeric value of zero and can be used to indicate the end of a sequence of letters, such as a word or sentence. This aids programmers in determining string length. Null characters are utilized as space fillers in practical applications such as database and spreadsheet programs.

Num Lock

Num Lock is a toggle key that changes the numeric keypad's input. The keypad can be used to enter numeric values when Num Lock is turned on. For example, the keys provide different input when Num Lock is turned off. 4 is a left arrow, 6 is a right arrow, 8 is an up arrow, and 2 is a down arrow. For example: "To use the numeric keypad as arrow keys, turn Num Lock off."

The Num Lock is rarely utilized since most keyboards have arrow keys. However, the Num Lock feature is still available on some keyboards. If you're having trouble entering digits on the numeric keypad, you might be able to solve the problem by pressing the Num Lock key.

Nybble

A nybble, sometimes known as a "nibble," is a four-bit set. Because a byte is made up of eight bits, a nybble is half of a byte. While it may take numerous nibbles to equal one bite of a cookie for the ordinary person, two nybbles always equal one byte in the computer world.

A nibble has 16 potential values thanks to its four bits, which is the same as one hexadecimal digit. As a result, a nybble is also known as a "hex digit." Because of the four bits that make up each nybble, nybbles are also referred to as "quadbits" in data transfers.

O

ODBC

Stands for "Open Database Connectivity." With so many various types of databases to choose from, such as Microsoft Access, Filemaker, and MySQL, it's critical to have a common manner of transferring data to and from each. The SQL Access group produced the ODBC standard in 1992 to address this issue. Any application that supports ODBC can access information from an ODBC-compatible database, regardless of what database management system the database uses.

An ODBC database driver is required for an ODBC database to be ODBC compatible. This enables other apps to connect to the database and access data using a standard set of commands. The driver converts conventional ODBC commands into commands that the database's proprietary technology can understand. ODBC allows a single application (such as a web server program) to access data from multiple databases with a single set of commands.

Offline

A computer or other device is "offline" if it is not turned on or connected to other devices. This is the opposite of "online," which refers to a device's ability to communicate with other devices. If you try to print to your printer and receive one of those annoying warnings that says "The specified printer could not be located," your printer is most likely offline. Make sure that the printer is connected properly and is turned on as well.

Offline can also refer to a lack of Internet access. Your computer goes offline when you disconnect from your ISP or disconnect the ethernet wire from your computer. The option to "Work Offline" is available in some programs, such as Web browsers and e-mail clients. This option stops the program's network connection, preventing data from being sent to or received from the computer. When most people utilized dial-up connections, this option was more useful. They didn't want their computer to dial their ISP every time a software attempted to connect to the Internet. However, most people today have "always on" connections such as DSL and cable modems, so working offline is rarely necessary.

OLAP

Stands for "Online Analytical Processing." Users can use OLAP to evaluate data from many database systems at the same time. While relational databases are thought to be two-dimensional, OLAP data is multidimensional, which means it may be compared in a variety of ways. For instance, a company might compare their computer sales in June with sales in July, then compare those results with the sales from another location, which might be stored in a different database.

To use OLAP to process database information, you'll need an OLAP server to arrange and compare the data. Clients can use the OLAP server's built-in functionalities to examine various sets of data. Oracle Express Server and Hyperion Solutions Essbase are two popular OLAP server software products. OLAP processing is frequently used for data mining, which tries to identify new links between disparate sets of data using its extensive data analytic skills.

OLE

Stands for "Object Linking and Embedding." It can be pronounced as "O-L-E," or "Oh-lay!" if you are feeling Spanish. OLE is a Microsoft foundation that allows you to take items from one application and place them in another. It was first introduced in Windows 3.1. For example, you might be able to use OLE to transfer an image from a photo editing program to a word processing document.

Object linking between "compound documents," or documents that support several sorts of data, was the original goal of the OLE technology. Since then, Microsoft has expanded OLE into a broader standard known as the Component Object Model (COM). Although COM is compatible with Mac, Unix, and Windows platforms, it is most commonly utilized with Microsoft Windows. The COM framework is the foundation of ActiveX, which allows developers to create interactive content for the Web.

Online

A machine is "online" when it is switched on and connects to other devices. When a network printer is online, for example, computers on that network can print from it. When scanners, video cameras, audio interfaces, and other devices are running and connected to a computer system, they are said to be online.

However, recently, "online" mainly refers to being connected to the Internet. A phone line with a dial-up or DSL modem, a cable line with a cable modem, or a wireless connection can all be used to connect. A computer can also be connected to a computer network and so be online. Even if they are not linked to the Internet, machines on a network are technically online. Most networks, however, are already routed to a T1 line or other Internet connection. A computer or other device is said to be offline when it is not connected to the internet.

OOP

Stands for object-oriented programming and refers to a programming methodology based on objects, instead of just functions and procedures. Individual objects can be grouped together thanks to the organization of these things into classes. Object-oriented programming languages include Java, C/C++, and PHP, as well as many earlier programming languages that now have object-oriented variants.

A specific type, or "instance," of a class is referred to as an "object" in an OOP language. Each object in the class has a structure that is identical to the other objects in the class, but it can be given unique properties. An object can also call functions or methods that are unique to it. A class that defines the structure of characters in a video game, for example, may be included in the source code of the game. Characters can be designed as objects, allowing them to have a variety of appearances, talents, and abilities. They can also carry out various jobs in the game, which are carried out utilizing the procedures relevant to each object.

Object-oriented programming simplifies the structure and organization of software programs for programmers. It is also easier to update and change programs written in object-oriented languages since individual objects can be changed without affecting other components of the program. OOP has made designing massive software programs more manageable as software programs have become larger over time.

Opacity

Opacity (pronounced "o-PASS-ity") refers to the opaqueness of an object. In computing, opacity is often used in graphics software to define how "non transparent" an image is. In other words, the more transparent an image or image layer is, the lower its opacity.

You may modify the opacity of an image layer with an Opacity slider in Adobe Photoshop and other graphics editors. This option usually has a range of 0 to 100. The opacity of a layer can be set to 0 to make it fully transparent or invisible. When you set it to 50, it becomes half transparent, or "see-through." If you give a layer an opacity of 100 (which is normally the default), it will be completely transparent. In a multi-layer image, all layers underneath a layer that fills the canvas with an 100% opacity will be completely hidden.

Open Firmware

Open Firmware is a type of firmware that is used by some computer systems to start up. Before the operating system is loaded, it regulates the CPU and performs system diagnostics. Open Firmware also creates the "device tree," which organizes all of the computer's internal and external devices. Each device is then given its own address, allowing it to be used after the computer has booted up.

Open Firmware is used on a variety of computers, including Macintosh PowerPC systems, Sun Microsystems SPARC-based workstations, and IBM POWER systems. (The BIOS serves the similar job on most Windows-based PCs.) Devices that support Open Firmware can often be utilized in multiple Open Firmware-based systems because it is an "open" standard. For example, identical PCI cards could be used in both Sun and Macintosh-based computer systems.

Press and hold "Command-Option-O-F" during start-up on a PowerPC-based Macintosh to access the Open Firmware interface. The Open Firmware interface appears at start-up on Sun systems and can be accessed at any time by hitting "L1-A" (or Stop-A) while the computer is operating.

Open Source

The term "open source" refers to a software program whose source code is freely available to the public. Open-source software, unlike commercial software, can be modified and distributed by anyone, and it is frequently produced by a community rather than a single business. As a result, the term "open-source community" is frequently used to characterize open-source software development projects.

Because anyone can modify the source code of an open-source program, it only makes sense that the software is also free to download and use. The GNU General Public License, which acts as the software license agreement

(SLA) for many open-source programs, is frequently used to establish the terms of use. Donors with an interest in the project, user donations, and marketing are all common sources of funding for open-source software development initiatives. Some software developers also make money by selling documentation and help manuals. Other projects are supported only by the desire of developers to make a fantastic program.

OpenGL

OpenGL, or the Open Graphics Library, is a 3D graphics language developed by Silicon Graphics. Prior to the release of OpenGL, software developers had to create custom 3D graphics code for each operating system platform and graphics hardware. OpenGL, on the other hand, allows developers to create visuals and special effects that will look nearly identical on any operating system and hardware that supports the technology. This makes it much easier for developers of 3D games and programs to port their software to multiple platforms.

So how exactly does OpenGL work? When writing OpenGL code, programmers specify a series of commands. Each command performs a drawing activity or generates a unique effect. Programmers can create 3D landscapes with unique effects like texture mapping, transparency (alpha blending), hidden surface removal, antialiasing, fog, and lighting effects by using hundreds or even thousands of these OpenGL instructions. An unlimited amount of viewing and modelling transformations can be applied to the OpenGL objects, giving developers an infinite number of possibilities.

Operating System

Also referred to as "OS," this is the software that communicates with computer hardware on the most basic level. Without an operating system, no software programs can run. The OS is what allocates memory, processes tasks, accesses disks and peripherals, and serves as the user interface.

Developers can use a standard programming interface, or API, to write code on operating systems like Windows, Mac OS, and Linux. To achieve the same outcomes without an operating system, programmers would have to write ten times as much code. of course, the operating system must be programmed by some computer wizards.

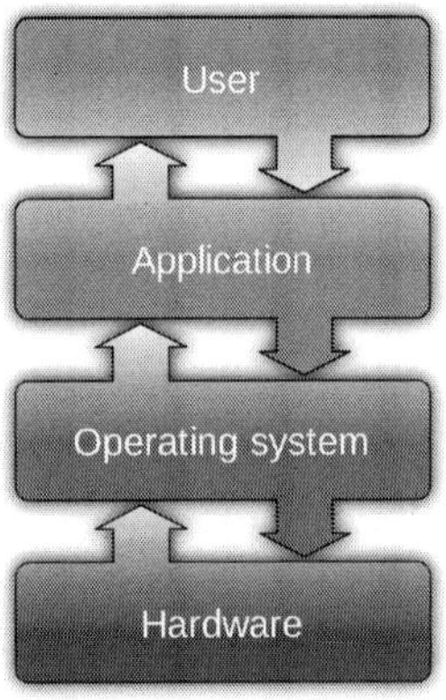

Figure 70. Operating system placement.

Source: Image by Wikimedia.

Optical Drive

"Optical" relates to eyesight, or the ability to see, in the real world. Lasers, which can "see" and read data on optical discs, are referred to as "optical" in the computer world. CDs and DVDs, for example, are made up of millions of little bumps and dips. Lasers in optical drives convert these bumps and dips into ones and zeros that the computer can understand.

CD-ROM, CD-RW, DVD-ROM, DVD-RW, and Blu-ray drives are all common types of optical drives. A laser is used in CD and DVD writers, such as CD-R and DVD-R drives, to read and write data on the discs. Because it "burns" the bumps and dips into the disc, the laser used to write the data is significantly more powerful than the laser needed to read the data. While optical drives may spin discs at high speeds, they are still far slower than hard drives, which use magnetic storage to store data. Optical media, on the other hand, is the most used medium for distributing computer software because it is inexpensive and removable.

OSD

OSD is short for "On Screen Display." An OSD is an onscreen menu found on most monitors that allows users to change the display's settings. Brightness, contrast, and color calibration adjustments are all common OSD settings. Positioning and tilt control are also available on some monitors. The OSD can be accessed by pushing the MENU button on the monitor's side. Once the OSD appears, you can navigate through the menu options and adjust plus (+) and minus (–) buttons. For example, you can adjust the brightness of your display using your monitor's OSD.

OSPF

Stands for "Open Shortest Path First." It is a method of finding the shortest path from one router to another in a local area network (LAN). The OSPF algorithm will calculate the most efficient way for data to be transmitted if the network is IP-based.

If a network has multiple routers, OSPF creates a table (or topography) of the router connections. When data is transported from one site to another, the OSPF algorithm weighs the choices and selects the most efficient method of transmission. This eliminates excessive data transfer delays and avoids infinite loops.

Outbox

Outgoing e-mail messages are temporarily stored in an outbox. Most email programs automatically save a draught of your message in the outbox as you're writing it. The message is then saved in the outbox until it is delivered to the intended recipient. Most e-mail programs move the message to the "Sent" or "Sent Messages" folder once it has been sent. Although the terms "Outbox" and "Sent Messages" are frequently used interchangeably, they have distinct meanings.

Unlike the inbox, which is frequently packed with e-mail, the outbox is frequently empty. Because all messages that have been sent have already been moved to the Sent Messages folder, this is the case. You can think of an e-mail outbox in the same way you would an office outbox. Mail that has to be delivered is temporarily stored in the outbox until it is picked up by the mailman (or the designated office mail person) and brought to the post office. Messages in an e-mail outbox, on the other hand, are usually dispatched right away (unless a connection to the outgoing SMTP mail server is not available).

Output

Output refers to the data produced by a computer. Data produced at a software level, such as the result of a calculation, or at a physical level, such as a printed document, falls under this category. A calculator program that produces the result of a mathematical operation is a fundamental example of software output. A more complex example is the results produced by a search engine, which compares keywords to millions of pages in its Web page index.

Output devices are devices that allow the computer to generate physical output. The monitor, which shows data on a screen, is the most often used output device. Other often used output devices include printers and computer speakers.

The opposite of output is input, which is data that is entered into the computer. Input and output devices are collectively referred to as I/O devices.

Overwrite

In computing, overwriting refers to replacing old data with new data. There are two primary types of overwriting: 1) replacing text, and 2) replacing files.

1) Replacing text

Most word processing systems insert characters where the cursor is currently placed by default. Some programs, on the other hand, allow you to modify the default behavior from insert to overwrite (or "overtype"). When an application offers both insert and overwrite modes, the Insert (INS) key is frequently used to switch between them. As new text is input in insert mode, the text to the right of the cursor is relocated to the right. For example, in the string "four, five, six," you might want to include the word "and" between "five" and "six." You'd place your cursor just before the word "six," then type "and" (followed by a space). "Four, five, and six" would be the result. The word "six" would be overwritten by the word "and," in the overwrite mode, so the resulting string would read, "four, five, and." Overwrite mode simply replaces existing characters as you type.

2) Replacing files

Overwriting also refers to the process of replacing old files with fresh ones. You may be asked if you wish to overwrite an existing document if you try to save a document with the same filename. The previous document will be overwritten by the new one if you click OK. When you move files to a folder, the operating system may ask if you want to overwrite existing files with the identical filenames. If you select Overwrite, the old files will be overwritten with the new ones.

P

Packet

This is a small bit of computer data that is transferred across the internet. When you get data from the Internet, it arrives in your computer as a series of little packets. Each packet contains the origin and destination addresses, as well as information that links it to the other packets being transmitted. "Packet-switching" is the process of sending and receiving packets.

Multiple places can send packets over the same lines, which are then sorted and directed to different routes by various computers along the way. It functions similarly to the post office, except that billions of packets are transmitted every day and most arrive in less than a few seconds. Even FedEx same-day delivery pales in comparison.

Page Orientation

The way a document is printed or shown on a screen is referred to as page orientation. "Portrait," in which the height is more than the width, and "Landscape," in which the width is greater than the height, are the two basic types. The default setting on most printers is portrait mode, which means the page is printed vertically from top to bottom. Some programs allow the user to change the orientation to landscape mode within the "Page Setup" window. For example: "He changed the page orientation from portrait to landscape when creating the sign."

Page View

A page view is the number of times a user accesses a web page. Website monitoring software track page views, also called "pageviews," to record a website's traffic. The higher the number of page views on a website, the more traffic it receives. However, because each time a Web page is loaded, a page view is logged, a single user might accumulate a large number of page views on a single website. As a result, unique page views are frequently recorded to keep track of how many different visitors a website receives over time.

The terms "page views" and "website hits" are frequently interchanged. While the term "hit" is commonly used to describe a page view, a hit is officially logged for each object that loads during a page view. A single page visit, for example, will record four hits if it comprises HTML, two pictures, and a JavaScript reference. One-page view will record over two hundred photographs if a page includes over two hundred images.

Page views are more similar to impressions, which are used by online advertisers to track their performance. If one advertisement is displayed on each page, page views and impressions may be identical. When many ads are placed on each page, however, the number of ad impressions exceeds the number of page views.

Parallel Port

This interface, which can be found on the back of older PCs, is used to connect external devices like printers and scanners. It has a 25-pin connector (DB-25) and is larger than most modern interfaces. Because Centronics established the initial parallel port standard, the parallel port is also referred to as a Centronics interface. Because the printer is the most popular device connected to the parallel port, it is often known as a printer port.

The Enhanced Parallel Port is the most recent parallel port standard, and it supports the same connectors as the Centronics interface (EPP). This standard allows for bidirectional communication and can transport data ten times faster than the Centronics interface. However, since the parallel port is a rather dated technology, don't be startled if you see USB or Firewire interfaces completely replace parallel ports in the future.

Password

A password is a string of characters that is used to verify a user's identity on a computer system. You may, for example, have a computer account that requires you to log in. You must give a valid username and password to gain access to your account. A login is a term used to describe this combination. While usernames are normally available to the public, passwords are unique to each user.

The majority of passwords are made up of multiple characters, which can include letters, numbers, and most symbols, but not spaces. While it's important to create a password that's easy to remember, don't make it so easy to guess that others may guess it. The safest passwords are made up of a combination of letters and numbers rather than actual words.

Figure 71. Passwords are used for the security purpose.

Source: Image by Pixabay.

You can paste data into a computer document in the same way you can paste a note on a page of paper. The paste function allows you to paste copied data into text documents, pictures, Web browser address boxes, and almost any other location where data may be entered. To paste data, however, you must first copy it to the "Clipboard," a temporary storage area in your system's memory, or RAM. This is accomplished by first selecting the data you wish to copy and then selecting "Copy" from the Edit menu of the program.

You can paste data copied to the Clipboard into the same document or a different document in the same program once it has been copied to the Clipboard. You can even paste copied data into a separate program's page. However, you can usually only paste data into a document that contains data of the same type. For instance, you can't paste a picture or an audio file into the address area of your Web browser.

To paste a copied piece of data, go to the Edit menu in the program you want to paste it in and pick "Paste." Data is usually pasted wherever the flashing cursor is in text documents or programs with text fields. To paste the data, use the keyboard shortcut "Control-V" for Windows or "Command-V" for Mac OS. Because the "P" key is normally reserved for the "Print" shortcut, and "V" is immediately next to the "C" key, which is used for copying, the "V" key is used. So, while it may not seem sensible at first, it makes sense when you consider it.

Payload

Each unit of data transferred over the Internet contains both header information and the actual data being conveyed. The header specifies the packet's source and destination, whereas the payload refers to the actual contents. Because header information, also known as overhead data, is only utilized during transmission, it is removed from the packet once it arrives at its destination. Therefore, the payload is the only data received by the destination system.

PCI

Stands for "Peripheral Component Interconnect." It is a hardware bus designed by Intel and used in both PCs and Macs. Most add-on cards such as SCSI, Firewire, and USB controllers, use a PCI connection. Some graphics cards use PCI, but most new graphics cards connect to the AGP slot.

PCI slots are around 3.5" long and 0.5" high and are located at the back of your computer. So, before you run out and buy that Firewire extension card, double-check that you have at least one PCI slot.

PDF

Stands for "Portable Document Format." Adobe Systems created the PDF file format, which is a multi-platform file format. The text, fonts, graphics, and even formatting of documents from a number of apps are all captured in a PDF file. Even if he has a Mac and you have a PC, you can e-mail a PDF document to your friend and it will look the same on his screen as it does on yours. Since PDFs contain color-accurate information, they should also print the same way they look on your screen.

Peripheral

Any external device that offers input and output for the computer is referred to as a computer peripheral. Input peripherals, such as a keyboard and mouse, are input peripherals, and output peripherals, such as a monitor and printer, are output peripherals. Because they offer input and output for the computer, computer peripherals, or peripheral devices, are frequently referred to as "I/O devices." Some peripherals, such as external hard drives, provide both input and output for the computer.

Perl

Perl actually stands for "Practical Extraction and Report Language." It is a scripting language which uses a syntax similar to C/C++. Perl is commonly used by Web programmers to create scripts for Web servers. Perl is especially good at parsing text, so programmers often use it for reading and searching through text files.

Permalink

Short for "permanent link," a permalink is a URL that links to a specific news story or Web posting. Permalinks are most commonly used for blogs, which are frequently changed and updated. They give a specific Web address to each posting, allowing blog entries to be bookmarked by visitors or linked to from other websites.

Because most blogs are published on dynamic, database-driven Web sites, they do not come with a URL. A blog entry may appear on a user's home page, but it may not have its own Web page with the extensions ".html," ."asp," ."php," and so on. As a result, once the posting has become obsolete and is no longer visible on the front page, there may be no way to find it. Using a permalink to define the location of each posting prevents blog entries from fading off into oblivion.

PHP

Stands for "Hypertext Preprocessor." (It is a recursive acronym, if you can understand what that means.) PHP is an HTML-embedded Web scripting language. This means that PHP code can be put into a Web page's HTML. When a PHP page is accessed, the server that the page is hosted on reads or "parses" the PHP code. The output from the PHP functions on the website are normally returned as HTML code, which can be read by the browser. Users cannot see PHP code on a page because it is translated into HTML before the page is loaded. This makes PHP pages secure enough to access databases and other secure information.

PHP borrows a lot of its syntax from other languages like C, Java, and Perl. PHP, on the other hand, has a variety of unique features and capabilities. The language's purpose is to make it simple and quick for Web developers to create dynamically generated pages. PHP can also be used to build database-driven websites. The official site for PHP is PHP.net if you want to learn more about it.

Primary Key

A primary key is a database record's unique identifier. One of the fields of a table is usually designated as the primary key when it is constructed. The primary key is frequently a number, but it could also be a text field or another type of data. If a database contains definitions of computer terminology, for example, it would make sense if each phrase was only mentioned once. By defining the "Term" field as the primary key, it would ensure that no term is listed more than once in the database.

While the primary key of a table is normally associated with a single field, it can also be made up of many values. A database of news items, for example, might employ both the title and date columns to identify each entry. The main key would be made out of the "Title" and "Date" values, ensuring that no two entries have the same title on the same day.

Most database management systems (DBMS) allow fields to be defined as "Unique" if numerous fields in a table cannot have duplicate values. Setting a field as the primary key is comparable to this, but it's not the same. The primary key also serves as a pointer to the record, making it easy for the database management system to search, sort, and access the data. It also aids in the linking of records across several tables. As a result, while a table can have numerous "Unique" fields, each table can only have one primary key. For each record, the primary key field must be defined and not set to null. After all, an identifier that isn't defined isn't particularly useful.

Processor

The heart of a computer is this tiny chip. The processor, sometimes known as a "microprocessor," does all calculations such as adding, subtracting, multiplying, and dividing. The Intel Pentium chip is the most popular CPU in PCs, while the PowerPC chip is utilized in Macintosh computers (developed by Motorola, IBM, and Apple).

Megahertz, or cycles per second, is the unit of measurement for a computer's CPU speed. Higher megahertz, on the other hand, does not automatically imply better performance. Even while a 600-MHz chip has double the clock speed of a 300-MHz chip, this does not imply that the computer with the 600-MHz chip will run twice as quickly. This is because other factors such as the processor's efficiency, the bus design, the amount of memory available, and the software running on the computer all influence the machine's speed. Some processors may do more operations per clock cycle than others, resulting in higher efficiency than processors with faster clock speeds. This is why the PowerPC chip is typically faster in comparison to Pentium chips that are clocked at higher megahertz.

Protocol

When computers communicate with one another, each computer must adhere to the same set of rules and instructions. The term "protocol" refers to a set of communication regulations. There are many distinct protocols because there are so many different methods for computers to communicate with each other — far too numerous for the ordinary person to remember. Some examples of these different protocols include PPP, TCP/IP, SLIP, HTTP, and FTP. Can you guess what the last "P" in each acronym stands for? If you guessed "protocol," send yourself a congratulations e-mail.

Proxy Server

A proxy server is now used by almost all significant corporations, organizations, and universities. Before accessing information on the Internet, all computers on the local network must pass via this server. An organization can improve network performance and filter what people connected to the network can access by utilizing a proxy server.

A proxy server uses a caching technology to increase Internet connection speeds from a network. Caching caches recently seen Web pages, photos, and files on a local hard disc, eliminating the need to download them again from the Internet. A proxy server caches anything accessed through the network, unlike your Web browser, which may save recently seen stuff on your machine. That means that if Bob views a news report on CNN.com at 1:00 p.m. and Jill views the same page at 1:03 p.m., Jill will almost certainly obtain the page directly from the proxy server's cache. While this provides users with lightning-fast access to Web pages, it also implies that viewers may not be seeing the most recent update to each page.

A proxy server's other key function is to filter what is permitted into the network. A proxy server can filter HTTP, FTP, and Secure protocols, but HTTP is the most prevalent. The proxy server can control which websites users on the network have access to. Many businesses prefer to ban access to websites that contain unacceptable content, such as hacking information and pornography, but other websites can also be screened. If an employer discovers that employees are spending too much time on sites like eBay or Quicken.com, the proxy server can also ban those sites.

Pseudocode

Many software programs are written in a programming language such as C++ or Java. When writing computer source code, these languages have a specific syntax that must be followed. Pseudocode, on the other hand, is an informal means of describing a program rather than a programming language. It has no strict syntactic requirements and instead serves as a generic representation of a program's functions.

Understanding the code of numerous languages can be difficult because each programming language has its own grammatical structure. Pseudocode solves this difficulty by employing standard syntax and globally recognized English phrases. For example, a line of PHP code may read: if ($i < 10) { i++; }

This could be written in pseudocode as: if i is less than 10, increment i by 1.

By describing a program in pseudocode, programmers of all types of languages can understand the function of a program.

Because pseudocode is an informal language, it is mostly used to create a program overview or rough draught. Pseudocode cannot be turned into an executable program because it is not a programming language. As a result, if pseudocode is to become a usable application, it must be transformed into a specific programming language.

Python

Python is a popular programming language for building Web applications and software plug-ins. It has an uncluttered visual appearance and is designed to be extremely readable. Python is similar to other scripting languages such as Perl and PHP, except it makes more use of English words and has less punctuation. Programs written in Python can be saved as basic scripts (with a .py file extension) or compiled programs (with a .pyc extension).

Python is available for use from the terminal on practically all Unix-based platforms, including MacOS.

Figure 72. Python: a highly demanded programming language.

Source: Image by Wikimedia.

Q

QBE

Stands for "Query By Example." QBE is a user-friendly technique of conducting database queries that comes standard with many database programs. A user must typically create input commands using correct SQL (Structured Query Language) syntax without QBE. Almost all database programs support this standard language. However, if the syntax is slightly incorrect the query may return the wrong results or may not run at all.

Query By Example tool offers a user-friendly interface for entering queries. The user can specify the query she wishes to run by filling in blanks or selecting items rather than crafting a whole SQL command. For instance, a user might want to choose one entry from the table "Table1" with the ID 123. Using SQL, the user would need to input the command, "SELECT * FROM Table1 WHERE ID = 123." The QBE interface may allow the user to just click on Table1, type in "123" in the ID field and click "Search."

Most database packages have QBE, albeit the interface varies depending on the application. For example, Microsoft Access includes a totally graphical QBE interface called "Query Design View." Users can pick a query operator and fill in blanks with search terms using the phpMyAdmin application, which is utilized with MySQL. Whatever QBE implementation is provided with a program, the purpose is the same – to make it easier to run database queries and to avoid the frustrations of SQL errors.

Quad-Core

Four processing cores are found in quad-core CPUs. These cores function as distinct processors, yet they are all housed in the same chip. A quad-core CPU can execute instructions utilizing all four cores at the same time, but to do so successfully, the computer's software must support multiprocessing. For example: "Apple's Mac Pro workstations have quad-core CPUs."

Queue

To avoid any misunderstanding, this phrase is pronounced in the same way as the letter "Q." Some folks appear to think it's pronounced "kay" or "kyoo-ee," but it's not. A queue is a list of jobs that have been queued for processing. A job is simply added to the list of jobs when it is sent to a queue. Queues are frequently used in computer programs to organize tasks.

For example, when the CPU finishes one computation, it will process the next one in the queue.

A printer queue is a collection of documents awaiting printing. A document is sent to the printer queue when you decide to print it. The document will be printed right away if there are no other tasks in the queue. If there are existing tasks in the queue, the new document will be added to the list and printed once the others have been completed. Most printers nowadays come with software that allows you to sort, cancel, and add tasks to the printer queue manually. While this may not appear to be a significant benefit for the average home user, it can be useful in enterprises where multiple individuals use a single printer.

QWERTY

This term is used to describe a standard (Latin alphabet-based) keyboard. It is called so because the first six keys in the upper-left part of the keyboard spell out Q-W-E-R-T-Y. You can even call it a QWERTYUIOP keyboard, but since QWERTY is a lot easier, people prefer it more. In case you are wondering why the QWERTY keyboard is arranged like it is, the original reason was to reduce the jamming of typebars in typewriters as they moved to strike the paper.

August Dvorak designed a more efficient keyboard in 1932, putting the vowels and the five most common consonants in the middle row, with the goal of establishing an alternate rhythm between the right and left hands. Even if the keys on the Dvorak keyboard were more effectively placed, switching from the familiar QWERTY arrangement was and still is too much of a pain for most people. So, it appears like we'll have to make do with what we've got. Please accept my apologies, Dvorak.

R

RAM

Stands for "Random Access Memory," and is pronounced like the male sheep. It is made up of small memory chips that form a memory module and then these modules are installed in the RAM slots on the motherboard of your computer.

Every time you open an application, it is loaded into RAM from the hard disc. Because reading data from RAM is far faster than reading data from a hard drive, this is the case. The ability to run programs directly from the computer's RAM eliminates any lag time. The more RAM your computer has, the more data it can load from the hard drive into the RAM, allowing your computer to run faster. In fact, boosting RAM to your computer may be more useful to its performance than upgrading the CPU.

Open the "System" Control Panel to see how much RAM a Windows computer has. By right-clicking "My Computer" and selecting "Properties...," you may accomplish this. Select "About This Mac" from the Apple Menu to see how much RAM is installed in a Macintosh computer.

Refresh Rate

A "maximum refresh rate" is frequently included in the technical specifications of computer monitors. This value, which is expressed in hertz (Hz), indicates how many times the screen is repainted every second. CRT monitors typically have refresh rates of 60, 75, and 85 Hz. Over 100 Hz refresh rates are supported by some monitors.

The less visual flutter you see on the screen, the greater the refresh rate. Flicker is visible when the refresh rate is less than 60 Hz, which means you can know the screen is being updated instead of viewing a steady image. This flicker might be taxing on your eyes and lead them to weary fast if the refresh rate is too sluggish. As if sitting at a computer for several hours wasn't hard enough!

To avoid flicker, adjust your monitor's refresh rate to the highest feasible. This option can be found in Windows' Monitors control panel and Mac OS X's Displays system configuration. While 60 Hz is regarded a reasonable refresh rate, some users will discover that 85 Hz is far superior. The maximum refresh rate is determined by three factors: 1) The rate your video card supports, 2) the rate your monitor supports, and 3) the resolution your monitor is set at. Lower resolutions (i.e., 800x600) typically support higher refresh rates than higher resolutions (i.e., 1600x1200).

You may not be able to modify the refresh rate on an LCD monitor. Because most LCD monitors have a standard refresh rate that is considerably above the "flicker" point, this is the case. LCD displays create less flicker than CRT monitors because the pixels on an LCD screen stay light for longer before dramatically fading than CRT monitors.

Registry

This is a database that Microsoft Windows uses to keep information about the applications installed on a computer's settings. The desktop backdrop, program settings, and file extension affiliations are all part of this data.

Remote Access

As the name suggests, remote access is the ability to access your computer from a remote location. Users can operate faraway computers from their local workstation using programs like PC Anywhere (Windows), Remote Access (Mac), and Timbuktu (Windows and Mac). The remote client software must be installed on the local machine, and the remote server software must be installed on the remote machine, in order for a remote access connection to be established. Authenticating the connecting user nearly always necessitates the usage of a username and password.

Remote access is more than just being able to connect to a remote machine – it is the ability to control the machine once the connection has been made. A remote access program can turn your local computer into the remote computer to which you connect. This is ideal for folks who work from home on occasion, as well as server managers who need to update and replace their servers periodically. Most remote access systems also allow users to transfer files between local and faraway PCs, which can reduce commuting time significantly. While remote access is beneficial to many people, you should only allow it on your machine if you absolutely need it. You'll have to cope with yet another security issue as a result of this.

Resolution

Resolution can describe either how many pixels a monitor can display or how fine a printer can print.

1. *Monitors:* A small monitor's resolution can be 640 x 480, which implies there are 640 pixels horizontally and 480 pixels vertically across the screen. 800 x 600, 1,024 x 768, and 1,280 x 1,024 are some more typical monitor resolutions. The more pixels that can be displayed on the screen, the higher the resolution.
2. *Printers:* The resolution of a printer refers to how finely it can print. Dots per inch, or "dpi," is the unit of measurement. The higher the dpi, the more clear the image. The resolution of a scanner is also measured in dpi.

RFID

RFID is short for "Radio-Frequency Identification." Small integrated circuits that can be scanned with a radio transmitter are known as RFID tags. This implies that, unlike a credit card or a UPC label, they can simply be passed into the range of the transmitter rather than being swiped or scanned immediately. Inventory tracking, retail loss prevention, electronic toll booth payments, and keyless entry are just a few of the uses for these tags. For example: "RFID tags are often placed inside the retail packaging of small electronic devices."

RGB

Stands for "Red Green Blue." It refers to the three light colors (yes, red, green, and blue) that can be mixed to create any color. White light is formed when the greatest intensities of each color are combined. When each hue's intensity is set to zero, the result is black. The vibrant visuals you see on TVs and computer monitors are created using RGB. However, in print, the 4 colors – cyan, yellow, magenta, and black (CYMK) – are used to create color images.

ROM

Stands for "Read-Only Memory." Please don't get this term mixed up with RAM or a hard drive, as many people do. Before the system software loads, the computer employs ROM, which is memory carrying hardwired instructions. The instructions are read from a little program called the BIOS (Basic Input/Output System) in the ROM of PCs.

Root

The top level of a file system is referred to as root, or "root node." The root node is also known as the root directory since file systems are organized by directories. The root directory can be thought of as the top of an upside-down tree, with folders and files stretching beneath it. For example: "On a Unix system, system files can only be accessed by the root user."

The "root user," commonly known as "root," is a user account with administrative privileges on a computer system. The root user, unlike other users, has full access to all files and folders on the main hard disc. As a result, performing system updates and account alterations is frequently delegated to the root user of a Unix system.

Router

This is a physical device that sends data from a local area network (LAN) to another network connection (thus the name). A router works in the same way as a coin sorter, allowing only authorized machines to connect to other computer systems. Most routers additionally store log files of network activities on the local network.

Figure 73. A wireless router.

Source: Image by Flickr.

Runtime

A program is said to be in runtime while it is running or executing. The phrase is most commonly used by software developers to describe when a program has faults. A "runtime error" is an error that occurs during the execution of a program. A runtime error would be, for example, if a program told you that 2 Plus 2 equaled 5000. A memory leak occurs when a software consumes an excessive quantity of system memory. This is sometimes referred to as a runtime error. A compile-time error is another common sort of program error, in which the application fails to compile into an executable program. Syntax problems in computer code and file linking issues are examples of compile-time errors.

S

Safe Mode

Safe Mode is a feature of the Windows operating system that allows it to run with the bare minimum of system files. Because it uses a general VGA display driver rather than a vendor-specific driver, you'll probably only be able to operate with 16 colors and a 640x480 resolution. All third-party drivers for additional peripherals such as mouse, keyboards, printers, and scanners are also disabled under Safe Mode. Networking files and settings are not loaded in basic safe mode. This means that you won't be able to connect to the Internet or other computers on a network.

After an unexpected crash, Windows may not fully start, and the only way to get the machine to boot is to utilize Safe Mode. You can use a disc utility tool to restore corrupted files or folders on the hard drive once you've successfully booted the computer in Safe Mode. When you see a "Sync Out of Range" notice on your screen, you can also reboot into Safe Mode to see your display.

It's possible that your computer will become sluggish and annoyingly slow at times. Safe Mode will allow you to diagnose the issue and determine which files are causing the computer to slow down. When you contact technical support, you may be asked to boot into Safe Mode to begin troubleshooting. To boot your Windows computer into Safe Mode, hold down the F8 key while the computer is starting up. Then select Safe Mode from the list of boot options.

SATA

Stands for "Serial Advanced Technology Attachment," or "Serial ATA." It's a connector that connects ATA hard drives to the motherboard of a computer. SATA transfer rates start at 150MBps, which is substantially quicker than even the best ATA/100 drives, which are limited to 100MBps. Serial ATA is anticipated to replace the previous standard, Parallel ATA (PATA), which has been around since the 1980s for this and other reasons.

The SATA interface has various advantages over the PATA interface, in addition to quicker transfer speeds. For one thing, unlike Parallel ATA, SATA drives each have their own independent bus, so there is no competition for bandwidth. They also employ smaller, thinner cables, which improves ventilation within the computer. SATA cables can be up to one meter long, but PATA cables are limited to 40cm. This allows computer makers more freedom when it comes to designing the internal arrangement of their machines. Finally, Serial ATA only has seven wires, but Parallel ATA has 40. This means that electromagnetic interference with SATA devices is less likely.

In conclusion, Serial ATA is a superior and more efficient interface over the obsolete PATA standard. Make sure your computer has a SATA interface if you want to buy one that will handle fast hard drives for years to come.

Schema

A schema is a model or outline that specifies the organization of a specific type of data in computing. Schemas are most commonly used to describe XML files and databases, but they can also be used to define a variety of other data types.

An XML schema is a list of items that can be found in an XML file in a specific order. This structure assures that numerous connected XML documents have the same syntax and don't contain any unknown data. XML files can be easily parsed by a script or program if they follow a predefined schema.

Database schemas are frequently used to create a visual representation of a database's tables and information. Tables are represented by boxes, each of which contains a list of fields. Within the database, lines are used to connect tables and demonstrate how data flows between them. Database schemas often include comments that describe the tables and fields and may include labels that identify primary and unique keys within each table.

Script

A computer script is a set of instructions that a program or scripting engine follows. Scripts can be used to automate tasks on a local computer or to generate Web pages on the Internet. DOS scripts and VB Scripts, for example, can be used to automate tasks on Windows machines, whilst AppleScript scripts can be used to automate work on Macintosh computers. Web servers are frequently used to generate dynamic Web page content using ASP, JSP, and PHP scripts.

Script files are usually just text documents that contain instructions written in a certain scripting language. This means most scripts can be opened and edited using a basic text editor. However, when opened by the appropriate scripting engine, the commands within the script are executed. VB (Visual Basic) scripts, for example, will run when double-clicked, using Windows' built-in VB scripting support. Since VB scripts can access and modify local files, you should never run a VB script that you receive as an unknown e-mail attachment.

Server

A server, as the name suggests, provides information to computers that connect to it. When users connect to a server, they can access the server's applications, files, and other data. Web servers, mail servers, and LAN servers are all common types of servers. Several distinct server programs can be run on a single computer.

Snippet

A snippet is a brief piece of source code or text that can be pasted into a program or a Web page. HTML snippets frequently contain tables, forms, and other HTML data, whereas program snippets typically comprise simple functions. Some snippets may come packaged with software development tools, but they can also be written by the programmer. Snippets can save developers a lot of time by allowing them to reuse the same block of code in multiple places. For example: "The Web developer saved the form as an HTML snippet so he could use it in several different Web pages."

Spreadsheet

A spreadsheet is a type of document that uses a grid of horizontal rows and vertical columns to contain data. Numbers (1, 2, 3, etc.) are commonly used to identify rows, while letters are used to label columns (A, B, C, etc). Cells relate to individual row/column locations, such as C3 or B12. Each cell has the capacity to hold a single instance of data. Information can be recorded in a more structured fashion than plain text by entering data into a spreadsheet. The data can also be studied using formulas and calculations thanks to the row/column structure.

Each row of a spreadsheet, for example, may include information about a person who has an account with a specific company. Each column might record a distinct part of the person's data, such as their first and last names, address, phone number, favorite foods, and so on. The spreadsheet program may use this information to count the number of individuals who live in a specific zip code, create a list of everyone who likes fried veal, and conduct other computations. A spreadsheet is analogous to a database in this regard.

Spreadsheets, on the other hand, are more streamlined than databases and are particularly useful for dealing with numbers. Spreadsheets are widely utilized in scientific and financial applications because of this. A spreadsheet, for example, might record bank account data such as balances and interest rates. A column containing the account balances of numerous clients may be easily totalled to get the total value of all the clients' balances. To calculate the worth of the accounts in a year, multiply these amounts by the interest rate from another cell. After you've built the formula, changing the value of just the interest rate cell will change the anticipated value of all the other accounts.

SQL

Stands for "Structured Query Language," and can be pronounced as either "sequel" or "S-Q-L." It is a query language that is used to retrieve and modify data in a database. "Insert," "update," and "delete" are some popular SQL commands. SEQUEL stands for "Structured English Query Language," and it was first developed by IBM in 1975. It has experienced a number of changes since then, many of which are due to Oracle products.

SQL is now widely used for developing and managing Web databases. Despite the fact that SQL is currently regarded as a standard language, there are still a number of SQL variants, such as mSQL and mySQL. SQL commands can be run when a Web page loads using a scripting language like PHP. This enables the creation of dynamic Web pages that display different information each time they are loaded.

SRAM

Stands for "Static Random Access Memory." It might seem tempting to pronounce this term as "Sram," but it is correctly pronounced "S-ram." SRAM is a sort of RAM that stores data in a static manner, meaning that the data remains constant as long as the memory chip is powered. This is in contrast to DRAM (dynamic RAM), which stores data dynamically and requires refreshing the data stored in the memory on a regular basis. SRAM is faster and uses less power than DRAM because it stores data statically.

SRAM, on the other hand, is more expensive to produce than DRAM due to its more complex structure. Because of this complexity, the amount of data that can be stored on a single chip is limited, hence SRAM devices cannot hold as much data as DRAM chips. As a result, DRAM is frequently employed as the primary memory in personal computers. SRAM, on the other hand, is frequently employed in smaller applications like CPU cache memory and hard drive buffers. Other consumer electronics, ranging from major appliances to little children's toys, employ it as well.

Syntax

Each spoken language has a set of general principles for the construction of words and sentences. The language syntax is the collection of these rules. Syntax is used in computer programming languages to define how declarations, functions, instructions, and other statements should be organized.

The syntax of each computer programming language is different. Many languages have similar syntactic rules, whereas others have their own set of rules. C and Java, for example, have a very similar syntax, yet Perl has several properties that neither C nor Java have.

Regardless of the constraints, a program's source code must have the proper syntax in order to compile and run correctly. In reality, if the syntax is incorrect, the program will fail to compile and produce a "syntax error." Something as simple as a missing parenthesis or a forgotten semicolon can also be a syntax error at the end of a statement. Even these small errors will keep the source code from compiling.

Fortunately, most software development products come with a debugger that can help locate syntax mistakes in the source code. Consider a program with nearly 10,000 lines of code in which a syntax mistake is generated by a single missing semicolon. Manually locating the error would be preferable to finding a needle in a haystack. The debugger, of course, simplifies the debugging procedure for the programmer.

T

Tag

On clothes, tags usually indicate the brand, size of the garment, fabrics used, and the washing instructions. Tags on Web pages specify what should appear on the screen when the page loads. In HTML (hypertext markup language) and other markup languages, such as XML, tags are the most fundamental formatting tool. The <table> element, for example, is used to build a table on a Web page. The table's data is contained within the <table> tag, and the table is closed with the </table> tag.

If you want something to show up in bold on a Web page, you fundamentally use the bold tag. For example, the line:

This site is the <b>best website</b> ever!

would show up as: This site is the best website ever!

Since there is often a need to format content within more general tags, the tags can be "nested," meaning one tag can enclose one or more other tags. For example:

<font="Times">This is the Times font, and <i>this is in italics</i>. Pretty exciting, huh? </font>

would should up as: This is the Times font, and this is in italics. Pretty exciting, huh?

Tags are a basic component of HTML and are rather straightforward to comprehend. You can either develop your own Web site from scratch (using a text editor and inputting your own tags) or use a layout application like Macromedia Dreamweaver, which will generate the tags for you. Tags used in markup languages don't irritate or chafe your neck, which is a plus.

TCP/IP

Stands for "Transmission Control Protocol/Internet Protocol." The US military developed these two protocols in the early days of the Internet. The goal was to make it possible for computers to communicate over large distances. The TCP section is responsible for validating packet delivery. The IP element of the protocol refers to the transmission of data packets between nodes. Since then, TCP/IP has become the Internet's foundation. TCP/IP software is included in all major operating systems, including Unix, Windows, and Mac OS.

Telnet

This is an application that lets you use a text-based interface to log into a Unix computer. You can access specific resources on a Telnet server if you have a Telnet account, such as your home directory, e-mail account, FTP files, and so on. The disadvantage of Telnet is that it requires the usage of Unix commands, which can be difficult to learn.

Terabyte

A terabyte is 2 to the 40th power, or 1,099,511,627,776 bytes.

10 to the 12th power, or 1,000,000,000,000 bytes, is an estimate. A terabyte is 1,024 gigabytes in size, and it comes before the petabyte. Web servers and file servers may contain several terabytes of space, although consumer hard drives are normally measured in gigabytes. A 500GB hard disc is also referred to as a half-terabyte drive.

Text Editor

Any word processing tool that allows you to type and edit text is referred to as a text editor. It's not called a text editor for nothing, after all... Common text editors include WordPad and Notepad for Windows, as well as SimpleText and TextEdit for Mac. Larger programs like Microsoft Word and WordPerfect are text editors as well, but they come with a lot more functions.

Third Party

Third-party software refers to applications created by firms other than the operating system's creator. It could also refer to third-party plug-ins created by companies other than the original application's developer.

Computer hardware made by companies other than the original manufacturer is referred to as third-party hardware. Hard drives, video cards, memory modules, and peripheral devices are examples of third-party components. For example: "Several third-party disk utilities are available for both Mac and Windows computers."

TIFF

Stands for "Tagged Image File Format." It is a graphics file format that was developed in the 1980s to serve as the universal image format for a variety of computer platforms. Color depths ranging from 1-bit to 24-bit are supported by the TIFF format. Since the introduction of the original TIFF standard, several incremental enhancements have been made to the format, resulting in around 50 different TIFF formats. So much for the concept of a universal format. Because of its modest file size and Internet compatibility, JPEG has recently become the most preferred global format.

U

UDP

Stands for "User Datagram Protocol." It's a data-transfer protocol that's part of the TCP/IP stack. UDP is a "stateless" protocol, which means it does not recognize that the packets it is sending have been received. As a result, the UDP protocol is commonly used for streaming video. While you may see video skips or hear some fuzz in audio recordings, UDP transmission keeps the playback from totally halting.

UML

Stands for "Unified Modeling Language." This is a programming language for the creation of object-oriented software. Programmers frequently establish "objects," which are collections of structured data within programs, to better organize program code. UML was created for this purpose and has been standardized by the Object Management Group (OMG). The language has gained enough support that it has become a standard language for visualizing and constructing software programs.

Undo

Most programs provide an Undo command in the Edit menu, with the key "Ctrl+Z" (PC) or "Cmd-Z" (Mac). It's used to reverse the most recent action in a program. Typing or deleting text in a word processing program, drawing, or moving images in an image editor, and trimming media in a video or audio editing program are all examples of events that can be undone. Most actions can be immediately reverted by selecting "Undo." Many programs also enable "multiple undo," which allows you to undo multiple actions at the same time. For example: "If you accidentally delete a section of text, select 'Undo' to bring it back."

Unix

UNIX is another name for it, albeit the letters don't stand for anything. Bell Labs was the first to develop the Unix operating system in the 1960s. It became popular for high-level computing in the 1970s, but not for consumer computing. Unix achieved enormous popularity in the 1990s as a result of the fact that many Internet services were originally housed on

Unix workstations. It still leads the industry as the most common operating system for Web servers.

Even so, because there are so many various versions of Unix, it remains a bit of a muddled operating system. Ultrix, Xenix, Linux, and GNU are just a few examples, all of which run on a variety of hardware platforms, further complicating things. Most people will never need to use Unix, but computer nerds seem to have a strong desire to do so.

UPS

Stands for "Uninterruptible Power Supply." In the technology world, UPS is not just limited to a brown shipping company, it is also a type of power supply that uses battery backup to maintain power during unexpected power outages.

A surge protector with a high-capacity rechargeable battery is a typical consumer UPS. Smaller UPSs resemble bulky power strips, whereas larger UPSs can stand upright and resemble mini computers. Uninterruptible power supply (UPS) is used by many firms to keep their equipment working in the event of a power outage. While a UPS may only keep a computer operating for 15 minutes after the power goes off, this is usually enough time to save all your important documents and shut down the computer correctly. If you're working on a critical document or project that hasn't been saved recently, that extra time can be vital.

URL

Stands for "Uniform Resource Locator." A URL is the Internet address of a certain Web site or file. It employs forward slashes to designate separate folders and cannot contain spaces or certain other characters. http://www.cnet.com/, http://web.mit.edu/, and ftp:/info.apple.com/ are some examples of URLs. As you can see, not all URLs start with the letter "http." The first portion of a URL identifies the type of resource it refers to.

User Interface

A user interface is the way by which a user interacts with a piece of software or hardware. A software interface, for example, may consist of windows, icons, menus, and buttons that enable the user to interact with the application. A graphical user interface, or GUI, is another name for this. A remote control or a video game controller are examples of hardware interfaces. It could also refer to a camcorder, digital camera, or iPod's controls. The majority of today's user interfaces are created with a combination of hardware and software. For example: "The Apple iPhone has a touchscreen user interface."

V

VCI

Stands for "Virtual Channel Identifier." The VCI (virtual path indicator) is used in conjunction with the VPI (virtual path indicator) to determine where an ATM cell will go over a network. Many ISPs (Internet Service Providers) employ ATM, or asynchronous transfer mode, to transfer data to client computers. Because ATMs deliver packets via fixed channels, data sent over the conventional TCP/IP protocol is more difficult to follow.

Each ATM cell's VCI specifies the fixed channel over which the data packet should be delivered. In comparison to the VPI, which is just 8 bits, it is a 16-bit field. Because this numerical tag identifies the virtual channel to which each packet belongs, it prevents other data from interfering with it as it travels through the network.

Vector

A vector is a quantity characterized by both magnitude and direction in mathematics. A vector, for example, could be represented by a 1-inch arrow pointing at a 30 degree angle. Another vector might be 2.5 inches long and angled at 160 degrees. Vectors are used to specify routes in specific types of pictures, such as EPS files and Adobe Illustrator projects, in the computer world. These visuals are sometimes referred to as vector graphics since they are made up of vectors, or routes, rather than dots. Vector drawings may be resized up and down without sacrificing quality.

In computer science, a vector may refer to a type of one-dimensional array. For example, a vector called "fibonacci" that stores the first six values of the Fibonacci sequence would be defined as follows:

fibonacci [0] = 0, fibonacci [1] = 1, fibonacci [2] = 1, fibonacci [3] = 2, fibonacci [4] = 3, fibonacci [5] = 5

Arrays and vectors are comparable, but vectors have their own memory management algorithms. Arrays are limited to the memory structure provided by the programming language in which they are written, which is usually referred to as a stack. Vectors have a more dynamic structure, sometimes referred to as a heap, which allows them to use memory more freely. While an array requires a fixed amount of memory, the memory utilized by a vector can grow or shrink as more or less elements are added or removed.

Virtual Memory

Memory is the hardware that allows your computer to run programs and load the operating system. It is made up of one or more RAM chips, each with several memory modules. A computer's real memory is constrained by the amount of RAM installed. 256MB, 512MB, and 1GB are the most common memory sizes.

Because your computer's RAM is limited, it's possible that you'll run out of memory if you run too many programs at once. This is where the concept of virtual memory comes into play. Virtual memory expands your computer's accessible memory by expanding the "address space," or memory locations where data can be stored. It accomplishes this by allocating more memory from hard disc space. However, because the hard disc is significantly slower than the RAM, virtual memory data must be transferred back to real memory before it can be used.

Virtual Reality

While the word "virtual" is typically overused in the computer world, it is aptly placed in the phrase "virtual reality." Virtual implies "being or resulting in essence or effect but not in actual fact, shape, or name," according to the American Heritage Dictionary. It can also refer to something that is "produced, mimicked, or carried out by a computer or computer network." As a result, virtual reality is best described as a computer-generated illusion of reality.

By donning specific glasses and headphones and connecting them to a computer system running the virtual reality program, a person can experience a virtual reality world. These technologies immerse the user in the virtual world's sights and sounds. Some virtual reality systems allow users to wear gloves equipped with electronic sensors that can be used to manipulate virtual objects. As the user moves their head or hands, the computer moves the virtual world accordingly in real-time.

Virtualization

Virtualization allows many operating systems to operate on a single computer. A Mac OS X computer, for example, can run Windows while a Windows computer can run Linux using virtualization software. Multiple operating systems can operate at the same time and even communicate with each other thanks to virtualization software. A virtual machine is a computer that runs an alternative operating system (OS) on top of the original OS via virtualization. For example: "Virtualization allows a Macintosh computer to run Mac OS X and Windows at the same time."

Virus

A computer virus, like a real virus, is something you don't want to get. Computer viruses are little programs or scripts that can harm your computer's performance. These malicious small programs can generate, move, and delete files, as well as waste your computer's memory and cause it to malfunction. Viruses can replicate, attach to programs, and spread across networks. In reality, the most frequent way to obtain a virus is to open an infected e-mail attachment.

We all know how difficult it is to get a computer to perform properly even when it is in good health, let alone when it has been infected with a virus. As a result, preventing an attack is preferable to attempting to heal one. There are a variety of antivirus apps that check incoming files for viruses before they can harm your computer. Norton AntiVirus, McAfee VirusScan, and Virex are a few of these apps. To avoid a virus attack, it is a good idea to have one of these apps installed on your computer. It's also critical to update your virus definitions file at least once a month so that your antivirus software can keep up with the current threats. It's similar to drinking an orange juice every morning in that it strengthens and protects your computer's immune system.

VoIP

Stands for "Voice Over Internet Protocol," and is often pronounced "voip." VoIP, or Voice over Internet Protocol, is essentially a telephone connection over the Internet. Instead of using analog telephone lines, the data is transferred digitally over the Internet Protocol (IP). This allows users to communicate great distances and across the globe without incurring long distance or international phone expenses.

You'll need a computer, an Internet connection, and VoIP software to use VoIP. A microphone, analog telephone adapter, or VoIP phone are also required. You can use a basic microphone and speaker setup with many VoIP apps. Others require VoIP phones, which look and function like conventional phone handsets but connect to your computer through USB. You can use normal phones with your computer thanks to analog telephone adapters. IP phones, which connect to a router via Ethernet or wirelessly, are another alternative. These phones are equipped with essential VoIP software and do not require the use of a computer.

Vonage is the largest provider of VoIP services, but there are several other companies that offer similar services. While Vonage charges a monthly service fee, programs like Skype and PeerMe allow users to connect to each other and talk for free. However, these free services may offer fewer connections, lower audio quality, and may be less reliable than paid services like Vonage.

W

WAN

Stands for "Wide Area Network." It resembles a Local Area Network (LAN), but it is much larger. WANs, unlike LANs, are not constrained to a single place. Many wide area networks use telephone lines, fiber optic cables, or satellite links to connect over great distances. They can also be made up of smaller, interconnected LANs. The Internet can be thought of as the world's largest WAN. You could even call the Internet a Super WAN BAM if you wanted to.

Web 2.0

Web 2.0 is a word that was coined in 2004 to describe the World Wide Web's second generation. The name "2.0" derives from the software industry, where incremental version numbers are assigned to new versions of software products. The new generation of the Web, like software, contains new features and capability that were not previously available. Web 2.0, on the other hand, does not relate to a single version of the Internet, but rather to a set of technological advancements.

Some examples of features considered to be part of Web 2.0 are listed below:

Blogs – also known as Web logs, these allow users to post thoughts and updates about their life on the Web.

Wikis – sites like Wikipedia and others enable users from around the world to add and update online content.

Social networking – sites like Facebook and MySpace allow users to build and customize their own profiles and communicate with friends.

Web applications – a broad range of new applications make it possible for users to run programs directly in a Web browser.

Web 2.0 technologies enable users to interact in ways that were previously unavailable. Websites have grown considerably more dynamic and interconnected, resulting in "online communities," and making information sharing on the Internet even easier. Because most Web 2.0 capabilities are available for free, sites like Wikipedia and Facebook have grown at breakneck speed. More features are added to the sites as they evolve, drawing on the technologies already in place. While Web 2.0 may be a fixed term for the current era of the Internet, the technology itself is still evolving and changing.

Web Browser

You're probably reading this in a browser right now. The program that individuals use to access the World Wide Web is known as a web browser, or simply "browser." It recognizes text, graphics, hypertext links, Javascript, and Java applets in HTML code. The browser displays a well formatted page after rendering the HTML code. Common browsers include Firefox, Chrome, or Safari.

Web Host

A web host is required to publish a website online. The web host saves all your website's pages and makes them accessible to computers with Internet access. The domain name, such as "sony.com," is associated with an IP address that identifies a particular computer. The IP address is found and the web site is loaded from your web host when someone types your domain name into their browser's address box.

Web hosting software, such as Apache, OS X Server, or Windows Server, can be installed on anywhere from one to thousands of computers. The majority of websites you visit are accessed through a "shared host," which is a single machine that can host hundreds of websites. A "dedicated host," which is a single machine that solely hosts one website, is frequently used by larger websites. Sites with a lot of traffic, like apple.com and microsoft.com, use multiple machines to host one website.

You'll need to join up for a "Web hosting service" if you wish to publish your own website. Because there are so many options, finding a decent Web host shouldn't be difficult. Simply ensure that the Web server you select provides excellent technical support and minimal or no downtime. Typically, you'll be charged a monthly price that changes based on how much disc space and bandwidth your site will consume. So, before you sign up for a Web hosting service, figure out how big your site will be and how much traffic you expect.

Web Page

The World Wide Web is made up of web pages. These papers are written in HTML (hypertext markup language), which your browser will translate.

Static and dynamic web pages are both possible. Each time a static page is browsed, it displays the same content. The content of dynamic pages might change each time they are accessed. Scripting languages like PHP, Perl, ASP, and JSP are commonly used to create these pages. The scripts in the pages call server functions that return data such as the date and time, as well as database information. All of the data is returned as HTML code, so all your browser has to do when it receives the page is translate the HTML.

Figure 74. Web Page.

Source: Image by Pixabay.

Website

A website, sometimes known as a web site, is distinct from a web page. Even though the two terms are frequently used interchangeably, they should not. So, what's the difference between the two? A website is a collection of web pages, to put it simply. Amazon.com, for example, is a website, yet it is made up of millions of web pages.

Wi-Fi

The Wi-Fi Alliance has trademarked Wi-Fi as a wireless networking protocol. All networking equipment that adheres to one of the IEEE 802.11 standards is referred to as IEEE 802.11. Wi-Fi enables computers and other devices to connect to wireless routers and, as a result, other network systems. Devices connected to the wireless access point may have Internet access if the router is connected to the Internet. Nowadays many coffee shops offer customers Internet access through a Wi-Fi connection.

X

X86

Intel processors released after the original 8086 processor are referred to as x86. The 286, 386, 486, and 586 processors are among them. The "x" in x86, as you can see, stands for a range of conceivable values. The complete names of the processors are 80286, 80386, 80486, and 80586, hence x86 is short for 80x86. To prevent repetition, the "80" is usually shortened.

When a computer's technical specifications say it's based on the x86 architecture, that suggests it's powered by an Intel CPU (not AMD or PowerPC). Newer x86 processors can run all of the programs that older processors could run because Intel's x86 chips are backwards compatible. Older CPUs, on the other hand, may be unable to run software designed for newer x86 processors.

XHTML

Stands for "Extensible Hypertext Markup Language." Yes, here "Extensible" starts with an "X." XHTML is a spinoff of the hypertext markup language (HTML) used for creating web pages. It is based on the HTML 4.0 syntax, but has been modified to follow the guidelines of XML, the Extensible Markup Language.

Web developers can construct their own objects and tags for each web page they make since XHTML is "extensible." Developers have more control over the appearance and arrangement of their web pages as a result of this. The only stipulation is that the custom tags and attributes be defined in a document type definition (DTD) that the XHTML page references.

In addition, XHTML pages must follow a stricter syntax than conventional HTML pages. While HTML syntax is more tolerant and forgiving in web browsers, XHTML documents must have precise syntax. This signifies the markup language has no missing quotes or erroneous capitalization. While the tight syntax necessitates more careful web page construction, it also assures that Web pages seem consistent across browser platforms.

XML

Stands for "Extensible Markup Language." (Yes, technically it should be EML). XML is a standard format for defining documents that may be read by any XML-compatible application. Although XML can be used with HTML pages, it is not a markup language in and of itself. It's a "metalanguage" that can be used to construct markup languages for certain applications. It can be used to define elements that can be accessed when a web page loads. Essentially, XML allows you to establish an information database without needing to use a database. XML documents are often used in web applications, but they can also be utilized in a variety of other programs.

Z

Zip

Using a tool like WinZip, you may archive and compress several files into a single smaller file. It's a convenient approach to compress files before sending them over email.

BIBLIOGRAPHY

1. Avi Networks. (2021). *What is Subnet Mask? Definition & FAQs | Avi Networks*. [online] Available at: https://avinetworks.com/glossary/subnet-mask/ (accessed 4 August 2021).
2. Biscobing, J., (2017). *What is OLAP (Online Analytical Processing)? – Definition from WhatIs.com*. [online] SearchDataManagement. Available at: https://searchdatamanagement.techtarget.com/definition/OLAP (accessed 4 August 2021).
3. Christensson, P., (2011). *File Type Definition*. [online] Techterms.com. Available at: https://techterms.com/definition/file_type (accessed 4 August 2021).
4. Christensson, P., (2014). *Fiber Optic Cable Definition*. [online] Techterms.com. Available at: https://techterms.com/definition/fiber_optic_cable (accessed 4 August 2021).
5. Christensson, P., (2015). *Gateway Definition*. [online] Techterms.com. Available at: https://techterms.com/definition/gateway (accessed 4 August 2021).
6. Dictionary.cambridge.org. (2021). *Bandwidth*. [online] Available at: https://dictionary.cambridge.org/dictionary/english/bandwidth (accessed 4 August 2021).
7. Encyclopedia Britannica. (n.d.). *Peripheral Device | Definition & Examples*. [online] Available at: https://www.britannica.com/technology/input-output-device (accessed 4 August 2021).
8. Fitzgibbons, L., (2019). *What is a Checksum? – Definition from WhatIs.com*. [online] SearchSecurity. Available at: https://searchsecurity.techtarget.com/definition/checksum (accessed 4 August 2021).

9. Gibb, R., (2019). *What is Latency? How Latency Works (with Examples).* [online] Articles for Developers Building High Performance Systems. Available at: https://blog.stackpath.com/latency/ (accessed 4 August 2021).
10. Java.com. (n.d.). *What is Java Technology and Why Do I Need It?* [online] Available at: https://java.com/en/download/help/whatis_java.html (accessed 4 August 2021).
11. Pc.net. (2009). *Glossary of Computer and Internet Terms.* [online] Available at: https://pc.net/glossary/ (accessed 4 August 2021).
12. SearchSoftwareQuality. (2007). *What is Runtime? – Definition from WhatIs.com.* [online] Available at: https://searchsoftwarequality.techtarget.com/definition/runtime (accessed 4 August 2021).
13. Teravainen, T., (2017). *What Is Adware? – Definition from WhatIs.com.* [online] SearchSecurity. Available at: https://searchsecurity.techtarget.com/definition/adware (accessed 4 August 2021).
14. Villinger, S., (2021). *What is RAM and Why Is It Important?* [online] Available at: https://www.avast.com/c-what-is-ram-memory (accessed 4 August 2021).
15. WhatIs.com. (2011). *What is OpenGL (Open Graphics Library)? – Definition from WhatIs.com.* [online] Available at: https://whatis.techtarget.com/definition/OpenGL-Open-Graphics-Library (accessed 4 August 2021).
16. Cloudflare. (n.d.). *What is a router?* [online] Available at: https://www.cloudflare.com/en-in/learning/network-layer/what-is-a-router/ (accessed 4 August 2021).
17. Cloudflare. (n.d.). *What is DNS?* [online] Available at: https://www.cloudflare.com/en-in/learning/dns/what-is-dns/ (accessed 4 August 2021).
18. Cole, A., (2020). *What is an Optical Drive & What does an Optical Drive do?* [online] Data recovery tips. Recover deleted files on Mac, Windows. Available at: https://www.cleverfiles.com/howto/what-is-optical-drive.html (accessed 4 August 2021).
19. Collinsdictionary.com. (n.d.). *Binary definition and meaning.* [online] Available at: https://www.collinsdictionary.com/dictionary/english/binary (accessed 4 August 2021).
20. Computerhope.com. (2017). *What is a Memory Module?* [online] Available at: https://www.computerhope.com/jargon/m/memory-

module.htm (accessed 4 August 2021).

21. Computerhope.com. (2019). *What is Safe Mode?* [online] Available at: https://www.computerhope.com/jargon/s/safemode.htm (accessed 4 August 2021).
22. Computerhope.com. (2020). *What is a Hard Drive?* [online] Available at: https://www.computerhope.com/jargon/h/harddriv.htm (accessed 4 August 2021).
23. Computerhope.com. (2020). *What is a Parallel Port?* [online] Available at: https://www.computerhope.com/jargon/p/paraport.htm (accessed 4 August 2021).
24. Computerhope.com. (2021). *What is a File Extension?* [online] Available at: https://www.computerhope.com/jargon/f/fileext.htm (accessed 4 August 2021).
25. Default. (n.d.). *Ad Hoc Network Definition | Ad Hoc Connection | Computer Networks | CompTIA*. [online] Available at: https://www.comptia.org/content/guides/what-is-an-ad-hoc-network (accessed 4 August 2021).
26. Definitionmeaning.com. (n.d.). *What Does QBE Mean? – Business – Definition and Meaning of QBE – Definition Meaning*. [online] Available at: http://definitionmeaning.com/Business/qbe (accessed 4 August 2021).
27. Definitions.net. (n.d.). *What Does Page Orientation Mean?* [online] Available at: https://www.definitions.net/definition/page+orientation (accessed 4 August 2021).
28. Encyclopedia Britannica. (n.d.). *Syntax | Grammar*. [online] Available at: https://www.britannica.com/topic/syntax (accessed 4 August 2021).
29. Nolledo, M., (2020). *What is Google Drive? A Guide to Navigating Google's File Storage Service and Collaboration Tools*. [online] Business Insider. Available at: https://www.businessinsider.in/tech/news/what-is-google-drive-a-guide-to-navigating-googles-file-storage-service-and-collaboration-tools/articleshow/77226585.cms (accessed 4 August 2021).
30. www.dictionary.com. (n.d.). *Definition of Hacker | Dictionary.com*. [online] Available at: https://www.dictionary.com/browse/hacker (accessed 4 August 2021).
31. Calbimonte, D., (2021). *SQL Definition*. [online] SQL Shack – articles about database auditing, server performance, data recovery, and more.

Available at: https://www.sqlshack.com/sql-definition/ (accessed 4 August 2021).

32. Chakraborty, K., (2021). *What is Firmware? – Definition from Techopedia*. [online] Techopedia.com. Available at: https://www.techopedia.com/definition/2137/firmware (accessed 4 August 2021).
33. Forcepoint. (n.d.). *What Is a Firewall?* [online] Available at: https://www.forcepoint.com/cyber-edu/firewall (accessed 4 August 2021).
34. Forcepoint. (n.d.). *What Is Malware?* [online] Available at: https://www.forcepoint.com/cyber-edu/malware (accessed 4 August 2021).
35. GeeksforGeeks. (2021). *What is Local Host? – GeeksforGeeks*. [online] Available at: https://www.geeksforgeeks.org/what-is-local-host/ (accessed 4 August 2021).
36. Jackins, T., (2021). *What is Remote Access? Connect to your Computer from Anywhere*. [online] Splashtop Inc. Available at: https://www.splashtop.com/what-is-remote-access (accessed 4 August 2021).
37. Merriam-webster.com. (n.d.). *Definition of CACHE*. [online] Available at: https://www.merriam-webster.com/dictionary/cache (accessed 4 August 2021).
38. NGINX. (n.d.). *What Is Load Balancing? How Load Balancers Work*. [online] Available at: https://www.nginx.com/resources/glossary/load-balancing/ (accessed 4 August 2021).
39. PCMAG. (n.d.). *Definition of Logic Gate*. [online] Available at: https://www.pcmag.com/encyclopedia/term/logic-gate (accessed 4 August 2021).
40. Redhat. (2021). *What Is Open Source?* [online] Available at: https://www.redhat.com/en/topics/open-source/what-is-open-source#:~:text=Open%20source%20is%20a%20term,code%20as%20they%20see%20fit. (accessed 4 August 2021).
41. Software Reviews, Opinions, and Tips – DNSstuff. (2019). *What Is Network Topology? Best Guide to Types & Diagrams – DNSstuff*. [online] Available at: https://www.dnsstuff.com/what-is-network-topology (accessed 4 August 2021).
42. STRATO AG. (n.d.). *What Is IMAP and How Can I Use It?* [online] Available at: https://www.strato.com/faq/en_us/product/what-is-imap-and-how-can-i-use-it/ (accessed 4 August 2021).

43. Techopedia.com. (n.d.). *What is Chipset? – Definition from Techopedia.* [online] Available at: https://www.techopedia.com/definition/2172/chipset (accessed 4 August 2021).

44. www.javatpoint.com. (n.d.). *What is MAC Address – Javatpoint.* [online] Available at: https://www.javatpoint.com/what-is-mac-address (accessed 4 August 2021).

45. www.javatpoint.com. (n.d.). *XML Full Form – Javatpoint.* [online] Available at: https://www.javatpoint.com/xml-full-form (accessed 4 August 2021).

46. Developer.mozilla.org. (n.d.). *What is the Difference Between Webpage, Website, Web Server, and Search Engine? – Learn Web Development | MDN.* [online] Available at: https://developer.mozilla.org/en-US/docs/Learn/Common_questions/Pages_sites_servers_and_search_engines (accessed 4 August 2021).

47. Techopedia.com. (n.d.). *What is An Integrated Circuit (IC)? – Definition from Techopedia.* [online] Available at: https://www.techopedia.com/definition/2366/integrated-circuit-ic (accessed 4 August 2021).

48. TheFreeDictionary.com. (n.d.). *Registries.* [online] Available at: https://www.thefreedictionary.com/registries (accessed 4 August 2021).

49. Toonen, E., (2020). *What Is a Permalink?* [online] Yoast. Available at: https://yoast.com/what-is-a-permalink/ (accessed 4 August 2021).

50. Vocabulary.com. (2021). *Node – Dictionary Definition.* [online] Available at: https://www.vocabulary.com/dictionary/node (accessed 4 August 2021).

51. W3schools.com. (2021). *What is JavaScript.* [online] Available at: https://www.w3schools.com/whatis/whatis_js.asp (accessed 4 August 2021).

52. Wordstream.com. (2021). *Meta Search Engine: What Is a Meta Search Engine?* [online] Available at: https://www.wordstream.com/meta-search-engine (accessed 4 August 2021).

Index

M

N

O

P